D1472152

Eureka Math
Grade 7
Modules 5 & 6

Special thanks go to the Gordon A. Cain Center and to the Department of Mathematics at Louisiana State University for their support in the development of *Eureka Math*.

Lesson 1: Chance Experiments

Classwork

Have you ever heard a weather forecaster say there is a 40% chance of rain tomorrow or a football referee tell a team there is a 50/50 chance of getting a heads on a coin toss to determine which team starts the game? These are probability statements. In this lesson, you are going to investigate probability and how likely it is that some events will occur.

Example 1: Spinner Game

Suppose you and your friend are about to play a game using the spinner shown here:

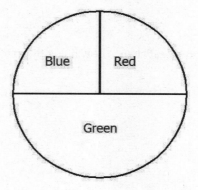

Rules of the game:

1. Decide who will go first.

2. Each person picks a color. Both players cannot pick the same color.

3. Each person takes a turn spinning the spinner and recording what color the spinner stops on. The winner is the person whose color is the first to happen 10 times.

Play the game, and remember to record the color the spinner stops on for each spin.

Exercises 1–4

1. Which color was the first to occur 10 times?

2. Do you think it makes a difference who goes first to pick a color?

3. Which color would you pick to give you the best chance of winning the game? Why would you pick that color?

4. Below are three different spinners. On which spinner is the green likely to win, unlikely to win, and equally likely to win?

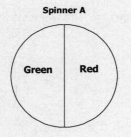

Spinner A
Green   Red

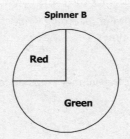

Spinner B
Red
Green

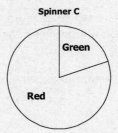

Spinner C
Green
Red

EUREKA
MATH™

Example 2: What Is Probability?

Probability is a measure of how likely it is that an event will happen. A probability is indicated by a number between 0 and 1. Some events are certain to happen, while others are impossible. In most cases, the probability of an event happening is somewhere between certain and impossible.

For example, consider a bag that contains only red cubes. If you were to select one cube from the bag, you are certain to pick a red one. We say that an event that is certain to happen has a probability of 1. If we were to reach into the same bag of cubes, it is impossible to select a yellow cube. An impossible event has a probability of 0.

Description	Example	Explanation
Some events are *impossible*. These events have a probability of 0.	You have a bag with two green cubes, and you select one at random. Selecting a blue cube is an impossible event.	There is no way to select a blue cube if there are no blue cubes in the bag.
Some events are *certain*. These events have a probability of 1.	You have a bag with two green cubes, and you select one at random. Selecting a green cube is a certain event.	You will always get a green cube if there are only green cubes in the bag.
Some events are classified as *equally likely to occur or to not occur*. These events have a probability of $\frac{1}{2}$.	You have a bag with one blue cube and one red cube, and you randomly pick one. Selecting a blue cube is equally likely to occur or not to occur.	Since exactly half of the bag is made up of blue cubes and exactly half of the bag comprises red cubes, there is a 50/50 chance (equally likely) of selecting a blue cube and a 50/50 chance (equally likely) of NOT selecting a blue cube.
Some events are more likely to occur than not to occur. These events have a probability that is greater than 0.5. These events could be described as *likely* to occur.	If you have a bag that contains eight blue cubes and two red cubes and you select one at random, it is likely that you will get a blue cube.	Even though it is not certain that you will get a blue cube, a blue cube would be selected most of the time because there are many more blue cubes than red cubes.
Some events are less likely to occur than not to occur. These events have a probability that is less than 0.5. These events could be described as *unlikely* to occur.	If you have a bag that contains eight blue cubes and two red cubes and you select one at random, it is unlikely that you will get a red cube.	Even though it is not impossible to get a red cube, a red cube would not be selected very often because there are many more blue cubes than red cubes.

Lesson 1:   Chance Experiments

The figure below shows the probability scale.

Probability Scale

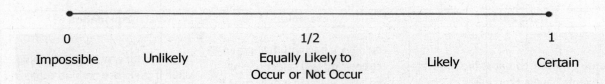

| 0 | | 1/2 | | 1 |
| Impossible | Unlikely | Equally Likely to Occur or Not Occur | Likely | Certain |

Exercises 5–10

5. Decide where each event would be located on the scale above. Place the letter for each event in the appropriate place on the probability scale.

Event:

A. You will see a live dinosaur on the way home from school today.

B. A solid rock dropped in the water will sink.

C. A round disk with one side red and the other side yellow will land yellow side up when flipped.

D. A spinner with four equal parts numbered 1–4 will land on the 4 on the next spin.

E. Your full name will be drawn when a full name is selected randomly from a bag containing the full names of all of the students in your class.

F. A red cube will be drawn when a cube is selected from a bag that has five blue cubes and five red cubes.

G. Tomorrow the temperature outside will be −250 degrees.

6. Design a spinner so that the probability of spinning a green is 1.

7. Design a spinner so that the probability of spinning a green is 0.

8. Design a spinner with two outcomes in which it is equally likely to land on the red and green parts.

An event that is impossible has a probability of 0 and will never occur, no matter how many observations you make. This means that in a long sequence of observations, it will occur 0% of the time. An event that is certain has a probability of 1 and will always occur. This means that in a long sequence of observations, it will occur 100% of the time.

9. What do you think it means for an event to have a probability of $\frac{1}{2}$?

10. What do you think it means for an event to have a probability of $\frac{1}{4}$?

Lesson Summary

- *Probability* is a measure of how likely it is that an event will happen.
- A probability is a number between 0 and 1.
- The probability scale is as follows:

Probability Scale

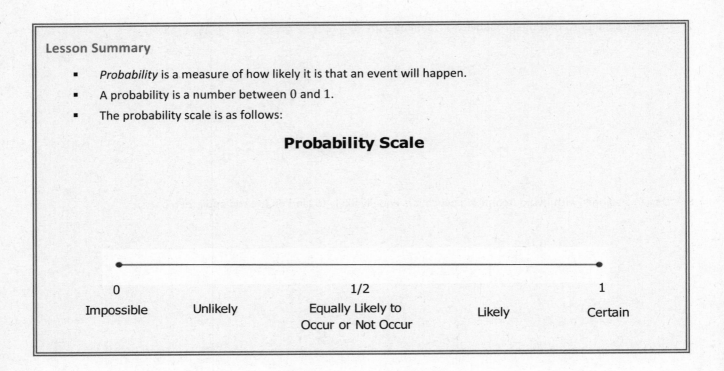

| 0 | | 1/2 | | 1 |
| Impossible | Unlikely | Equally Likely to Occur or Not Occur | Likely | Certain |

Problem Set

1. Match each spinner below with the words *impossible, unlikely, equally likely to occur or not occur, likely,* and *certain* to describe the chance of the spinner landing on black.

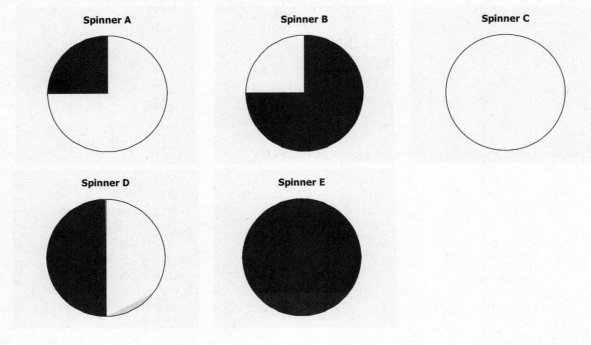

Spinner A Spinner B Spinner C

Spinner D Spinner E

2. Decide if each of the following events is *impossible, unlikely, equally likely to occur or not occur, likely,* or *certain* to occur.

    a. A vowel will be picked when a letter is randomly selected from the word *lieu*.

    b. A vowel will be picked when a letter is randomly selected from the word *math*.

    c. A blue cube will be drawn from a bag containing only five blue and five black cubes.

    d. A red cube will be drawn from a bag of 100 red cubes.

    e. A red cube will be drawn from a bag of 10 red and 90 blue cubes.

3. A shape will be randomly drawn from the box shown below. Decide where each event would be located on the probability scale. Then, place the letter for each event on the appropriate place on the probability scale.

Event:

A. A circle is drawn.

B. A square is drawn.

C. A star is drawn.

D. A shape that is not a square is drawn.

Probability Scale

0		1/2		1
Impossible	Unlikely	Equally Likely to Occur or Not Occur	Likely	Certain

4. Color the squares below so that it would be equally likely to choose a blue or yellow square.

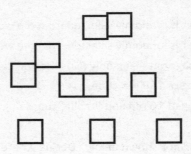

5. Color the squares below so that it would be likely but not certain to choose a blue square from the bag.

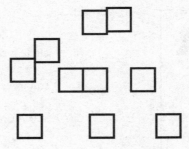

6. Color the squares below so that it would be unlikely but not impossible to choose a blue square from the bag.

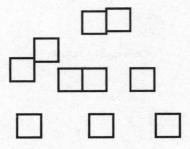

7. Color the squares below so that it would be impossible to choose a blue square from the bag.

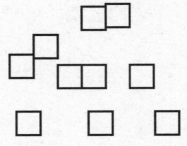

EUREKA
MATH™

Lesson 2: Estimating Probabilities by Collecting Data

Classwork

Exercises 1–8: Carnival Game

At the school carnival, there is a game in which students spin a large spinner. The spinner has four equal sections numbered 1–4 as shown below. To play the game, a student spins the spinner twice and adds the two numbers that the spinner lands on. If the sum is greater than or equal to 5, the student wins a prize.

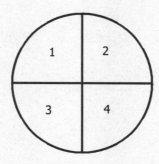

Play this game with your partner 15 times. Record the outcome of each spin in the table below.

Turn	First Spin Results	Second Spin Results	Sum
1			
2			
3			
4			
5			
6			
7			
8			
9			
10			
11			
12			
13			
14			
15			

1. Out of the 15 turns, how many times was the sum greater than or equal to 5?

2. What sum occurred most often?

3. What sum occurred least often?

4. If students were to play a lot of games, what fraction of the games would they win? Explain your answer.

5. Name a sum that would be impossible to get while playing the game.

6. What event is certain to occur while playing the game?

Lesson 2:   Estimating Probabilities by Collecting Data

EUREKA MATH™

When you were spinning the spinner and recording the outcomes, you were performing a *chance experiment*. You can use the results from a chance experiment to estimate the probability of an event. In Exercise 1, you spun the spinner 15 times and counted how many times the sum was greater than or equal to 5. An estimate for the probability of a sum greater than or equal to 5 is

$$P(\text{sum} \geq 5) = \frac{\text{Number of observed occurrences of the event}}{\text{Total number of observations}}.$$

7. Based on your experiment of playing the game, what is your estimate for the probability of getting a sum of 5 or more?

8. Based on your experiment of playing the game, what is your estimate for the probability of getting a sum of exactly 5?

Example 2: Animal Crackers

A student brought a very large jar of animal crackers to share with students in class. Rather than count and sort all the different types of crackers, the student randomly chose 20 crackers and found the following counts for the different types of animal crackers. Estimate the probability of selecting a zebra.

Animal	Number Selected
Lion	2
Camel	1
Monkey	4
Elephant	5
Zebra	3
Penguin	3
Tortoise	2
	Total 20

Exercises 9–15

If a student randomly selected a cracker from a large jar:

9.   What is your estimate for the probability of selecting a lion?

10. What is your estimate for the probability of selecting a monkey?

11. What is your estimate for the probability of selecting a penguin or a camel?

12. What is your estimate for the probability of selecting a rabbit?

13. Is there the same number of each kind of animal cracker in the jar? Explain your answer.

14. If the student randomly selected another 20 animal crackers, would the same results occur? Why or why not?

15. If there are 500 animal crackers in the jar, how many elephants are in the jar? Explain your answer.

EUREKA
MATH™

Lesson Summary

An estimate for finding the probability of an event occurring is

$$P(\text{event occurring}) = \frac{\text{Number of observed occurrences of the event}}{\text{Total number of observations}}.$$

Problem Set

1. Play a game using the two spinners below. Spin each spinner once, and then multiply the outcomes together. If the result is less than or equal to 8, you win the game. Play the game 15 times, and record your results in the table below. Then, answer the questions that follow.

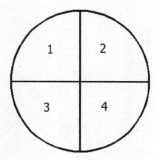 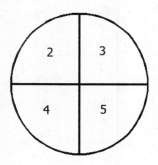

Turn	First Spin Results	Second Spin Results	Product
1			
2			
3			
4			
5			
6			
7			
8			
9			
10			
11			
12			
13			
14			
15			

    a.   What is your estimate for the probability of getting a product of 8 or less?

    b.   What is your estimate for the probability of getting a product of more than 8?

    c.   What is your estimate for the probability of getting a product of exactly 8?

    d.   What is the most likely product for this game?

    e.   If you play this game another 15 times, will you get the exact same results? Explain.

2.   A seventh-grade student surveyed students at her school. She asked them to name their favorite pets. Below is a bar graph showing the results of the survey.

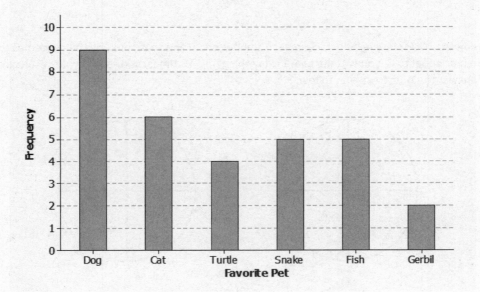

Use the results from the survey to answer the following questions.

    a.   How many students answered the survey question?

    b.   How many students said that a snake was their favorite pet?

Now, suppose a student is randomly selected and asked what his favorite pet is.

    c.   What is your estimate for the probability of that student saying that a dog is his favorite pet?

    d.   What is your estimate for the probability of that student saying that a gerbil is his favorite pet?

    e.   What is your estimate for the probability of that student saying that a frog is his favorite pet?

©2015 Great Minds. eureka-math.org
G7-M5-SE-B3-1.3.1-01.2016

3. A seventh-grade student surveyed 25 students at her school. She asked them how many hours a week they spend playing a sport or game outdoors. The results are listed in the table below.

Number of Hours	Tally	Frequency
0	| | |	3
1	| | | |	4
2	﹢﹢﹢﹢	5
3	﹢﹢﹢﹢ | |	7
4	| | |	3
5		0
6	| |	2
7		0
8	|	1

a. Draw a dot plot of the results.

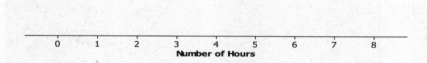

Suppose a student will be randomly selected.

b. What is your estimate for the probability of that student answering 3 hours?

c. What is your estimate for the probability of that student answering 8 hours?

d. What is your estimate for the probability of that student answering 6 or more hours?

e. What is your estimate for the probability of that student answering 3 or fewer hours?

f. If another 25 students were surveyed, do you think they would give the exact same results? Explain your answer.

g. If there are 200 students at the school, what is your estimate for the number of students who would say they play a sport or game outdoors 3 hours per week? Explain your answer.

4. A student played a game using one of the spinners below. The table shows the results of 15 spins. Which spinner did the student use? Give a reason for your answer.

Spin	Results
1	1
2	1
3	2
4	3
5	1
6	2
7	3
8	2
9	2
10	1
11	2
12	2
13	1
14	3
15	1

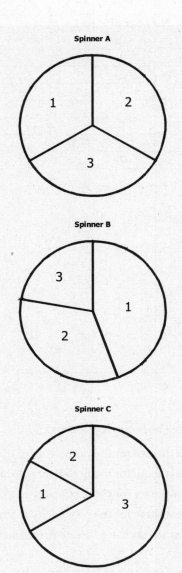

EUREKA
MATH™

Lesson 3: Chance Experiments with Equally Likely Outcomes

Classwork

Exercises 1–6

Jamal, a seventh grader, wants to design a game that involves tossing paper cups. Jamal tosses a paper cup five times and records the outcome of each toss. An *outcome* is the result of a single trial of an experiment.

Here are the results of each toss:

Jamal noted that the paper cup could land in one of three ways: on its side, right side up, or upside down. The collection of these three outcomes is called the *sample space* of the experiment. The *sample space* of an experiment is the set of all possible outcomes of that experiment.

For example, the sample space when flipping a coin is heads, tails.

The sample space when drawing a colored cube from a bag that has 3 red, 2 blue, 1 yellow, and 4 green cubes is red, blue, yellow, green.

For each of the following chance experiments, list the sample space (i.e., all the possible outcomes).

1. Drawing a colored cube from a bag with 2 green, 1 red, 10 blue, and 3 black

2. Tossing an empty soup can to see how it lands

3. Shooting a free throw in a basketball game

4. Rolling a number cube with the numbers 1–6 on its faces

5. Selecting a letter from the word *probability*

6. Spinning the spinner:

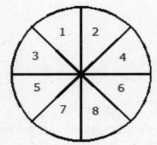

EUREKA
MATH™

Example 2: Equally Likely Outcomes

The sample space for the paper cup toss was on its side, right side up, and upside down.

The outcomes of an experiment are equally likely to occur when the probability of each outcome is equal.

Toss the paper cup 30 times, and record in a table the results of each toss.

Toss	Outcome
1	
2	
3	
4	
5	
6	
7	
8	
9	
10	
11	
12	
13	
14	
15	
16	
17	
18	
19	
20	
21	
22	
23	
24	
25	
26	
27	
28	
29	
30	

©2015 Great Minds. eureka-math.org
G7-M5-SE-B3-1.3.1-01.2016

Exercises 7–12

7. Using the results of your experiment, what is your estimate for the probability of a paper cup landing on its side?

8. Using the results of your experiment, what is your estimate for the probability of a paper cup landing upside down?

9. Using the results of your experiment, what is your estimate for the probability of a paper cup landing right side up?

10. Based on your results, do you think the three outcomes are equally likely to occur?

11. Using the spinner below, answer the following questions.

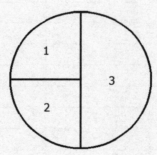

    a. Are the events spinning and landing on 1 or 2 equally likely?

EUREKA
MATH™

    b.   Are the events spinning and landing on 2 or 3 equally likely?

    c.   How many times do you predict the spinner will land on each section after 100 spins?

12. Draw a spinner that has 3 sections that are equally likely to occur when the spinner is spun. How many times do you think the spinner will land on each section after 100 spins?

Lesson Summary

An *outcome* is the result of a single observation of an experiment.

The *sample space* of an experiment is the set of all possible outcomes of that experiment.

The outcomes of an experiment are *equally likely* to occur when the probability of each outcome is equal.

Suppose a bag of crayons contains 10 green, 10 red, 10 yellow, 10 orange, and 10 purple crayons. If one crayon is selected from the bag and the color is noted, the *outcome* is the color that is chosen. The *sample space* will be the colors: green, red, yellow, orange, and purple. Each color is *equally likely* to be selected because each color has the same chance of being chosen.

Problem Set

1. For each of the following chance experiments, list the sample space (all the possible outcomes).

   a. Rolling a 4-sided die with the numbers 1–4 on the faces of the die

   b. Selecting a letter from the word *mathematics*

   c. Selecting a marble from a bag containing 50 black marbles and 45 orange marbles

   d. Selecting a number from the even numbers 2–14, including 2 and 14

   e. Spinning the spinner below:

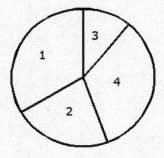

EUREKA
MATH™

2. For each of the following, decide if the two outcomes listed are equally likely to occur. Give a reason for your answer.

    a. Rolling a 1 or a 2 when a 6-sided number cube with the numbers 1–6 on the faces of the cube is rolled

    b. Selecting the letter *a* or *k* from the word *take*

    c. Selecting a black or an orange marble from a bag containing 50 black and 45 orange marbles

    d. Selecting a 4 or an 8 from the even numbers 2–14, including 2 and 14

    e. Landing on a 1 or a 3 when spinning the spinner below

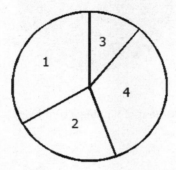

3. Color the squares below so that it would be equally likely to choose a blue or yellow square.

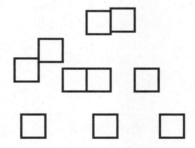

4. Color the squares below so that it would be more likely to choose a blue than a yellow square.

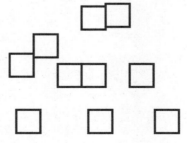

5. You are playing a game using the spinner below. The game requires that you spin the spinner twice. For example, one outcome could be yellow on the 1st spin and red on the 2nd spin. List the sample space (all the possible outcomes) for the two spins.

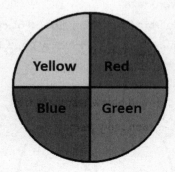

6. List the sample space for the chance experiment of flipping a coin twice.

EUREKA
MATH™

Lesson 4: Calculating Probabilities for Chance Experiments with Equally Likely Outcomes

Classwork

Examples: Theoretical Probability

In a previous lesson, you saw that to find an estimate of the probability of an event for a chance experiment you divide

$$P(\text{event}) = \frac{\text{Number of observed occurrences of the event}}{\text{Total number of observations}}.$$

Your teacher has a bag with some cubes colored yellow, green, blue, and red. The cubes are identical except for their color. Your teacher will conduct a chance experiment by randomly drawing a cube with replacement from the bag. Record the outcome of each draw in the table below.

Trial	Outcome
1	
2	
3	
4	
5	
6	
7	
8	
9	
10	
11	
12	
13	
14	
15	
16	
17	
18	
19	
20	

1. Based on the 20 trials, estimate for the probability of
    a. Choosing a yellow cube

    b. Choosing a green cube

    c. Choosing a red cube

    d. Choosing a blue cube

2. If there are 40 cubes in the bag, how many cubes of each color are in the bag? Explain.

3. If your teacher were to randomly draw another 20 cubes one at a time and with replacement from the bag, would you see exactly the same results? Explain.

Lesson 4: Calculating Probabilities for Chance Experiments with Equally Likely Outcomes

EUREKA MATH™

4. Find the fraction of each color of cubes in the bag.

      Yellow

      Green

      Red

      Blue

Each fraction is the *theoretical probability* of choosing a particular color of cube when a cube is randomly drawn from the bag.

When all the possible outcomes of an experiment are equally likely, the probability of each outcome is

$$P(\text{outcome}) = \frac{1}{\text{Number of possible outcomes}}.$$

An event is a collection of outcomes, and when the outcomes are equally likely, the theoretical probability of an event can be expressed as

$$P(\text{event}) = \frac{\text{Number of favorable outcomes}}{\text{Number of possible outcomes}}.$$

The theoretical probability of drawing a blue cube is

$$P(\text{blue}) = \frac{\text{Number of blue cubes}}{\text{Total number of cubes}} = \frac{10}{40}.$$

5. Is each color equally likely to be chosen? Explain your answer.

6. How do the theoretical probabilities of choosing each color from Exercise 4 compare to the experimental probabilities you found in Exercise 1?

7. An experiment consisted of flipping a nickel and a dime. The first step in finding the theoretical probability of obtaining a heads on the nickel and a heads on the dime is to list the sample space. For this experiment, complete the sample space below.

Nickel   Dime

What is the probability of flipping two heads?

Exercises 1–4

1. Consider a chance experiment of rolling a six-sided number cube with the numbers 1–6 on the faces.

   a. What is the sample space? List the probability of each outcome in the sample space.

   b. What is the probability of rolling an odd number?

   c. What is the probability of rolling a number less than 5?

Lesson 4:   Calculating Probabilities for Chance Experiments with Equally Likely
                        Outcomes   **EUREKA
                                                                                 MATH**™

©2015 Great Minds. eureka-math.org
G7-M5-SE-B3-1.3.1-01.2016

2. Consider an experiment of randomly selecting a letter from the word *number*.

   a. What is the sample space? List the probability of each outcome in the sample space.

   b. What is the probability of selecting a vowel?

   c. What is the probability of selecting the letter *z*?

3. Consider an experiment of randomly selecting a square from a bag of 10 squares.

   a. Color the squares below so that the probability of selecting a blue square is $\frac{1}{2}$.

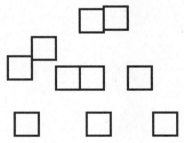

b.   Color the squares below so that the probability of selecting a blue square is $\frac{4}{5}$.

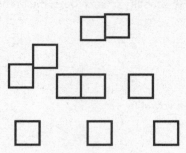

4.   Students are playing a game that requires spinning the two spinners shown below. A student wins the game if both spins land on red. What is the probability of winning the game? Remember to first list the sample space and the probability of each outcome in the sample space. There are eight possible outcomes to this chance experiment.

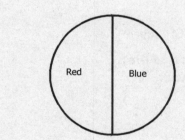 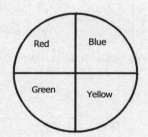

EUREKA
MATH™

Lesson Summary

When all the possible outcomes of an experiment are equally likely, the probability of each outcome is

$$P(\text{outcome}) = \frac{1}{\text{Number of possible outcomes}}.$$

An event a collection of outcomes, and when all outcomes are equally likely, the theoretical probability of an event can be expressed as

$$P(\text{event}) = \frac{\text{Number of favorable outcomes}}{\text{Number of possible outcomes}}.$$

Problem Set

1. In seventh-grade class of 28 students, there are 16 girls and 12 boys. If one student is randomly chosen to win a prize, what is the probability that a girl is chosen?

2. An experiment consists of spinning the spinner once.
    a. Find the probability of landing on a 2.
    b. Find the probability of landing on a 1.
    c. Is landing in each section of the spinner equally likely to occur? Explain.

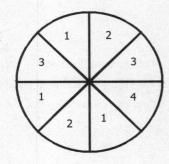

3. An experiment consists of randomly picking a square section from the board shown below.
    a. Find the probability of choosing a triangle.
    b. Find the probability of choosing a star.
    c. Find the probability of choosing an empty square.
    d. Find the probability of choosing a circle.

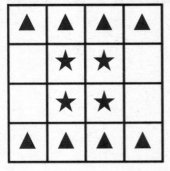

4. Seventh graders are playing a game where they randomly select two integers 0–9, inclusive, to form a two-digit number. The same integer might be selected twice.

    a. List the sample space for this chance experiment. List the probability of each outcome in the sample space.

    b. What is the probability that the number formed is between 90 and 99, inclusive?

    c. What is the probability that the number formed is evenly divisible by 5?

    d. What is the probability that the number formed is a factor of 64?

5. A chance experiment consists of flipping a coin and rolling a number cube with the numbers 1–6 on the faces of the cube.

    a. List the sample space of this chance experiment. List the probability of each outcome in the sample space.

    b. What is the probability of getting a heads on the coin and the number 3 on the number cube?

    c. What is the probability of getting a tails on the coin and an even number on the number cube?

6. A chance experiment consists of spinning the two spinners below.

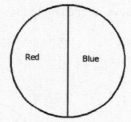 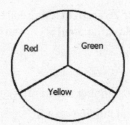

    a. List the sample space and the probability of each outcome.

    b. Find the probability of the event of getting a red on the first spinner and a red on the second spinner.

    c. Find the probability of a red on at least one of the spinners.

Lesson 5: Chance Experiments with Outcomes That Are Not Equally Likely

Classwork

In previous lessons, you learned that when the outcomes in a sample space are equally likely, the probability of an event is the number of outcomes in the event divided by the number of outcomes in the sample space. However, when the outcomes in the sample space are *not* equally likely, we need to take a different approach.

Example 1

When Jenna goes to the farmers' market, she usually buys bananas. The number of bananas she might buy and their probabilities are shown in the table below.

Number of Bananas	0	1	2	3	4	5
Probability	0.1	0.1	0.1	0.2	0.2	0.3

a. What is the probability that Jenna buys exactly 3 bananas?

b. What is the probability that Jenna does not buy any bananas?

c. What is the probability that Jenna buys more than 3 bananas?

d. What is the probability that Jenna buys at least 3 bananas?

e. What is the probability that Jenna does not buy exactly 3 bananas?

Notice that the sum of the probabilities in the table is one whole $(0.1 + 0.1 + 0.1 + 0.2 + 0.2 + 0.3 = 1)$. This is always true; when we add up the probabilities of all the possible outcomes, the result is always 1. So, taking 1 and subtracting the probability of the event gives us the probability of something *not* occurring.

Exercises 1–2

Jenna's husband, Rick, is concerned about his diet. On any given day, he eats 0, 1, 2, 3, or 4 servings of fruits and vegetables. The probabilities are given in the table below.

Number of Servings of Fruits and Vegetables	0	1	2	3	4
Probability	0.08	0.13	0.28	0.39	0.12

1. On a given day, find the probability that Rick eats

    a. Two servings of fruits and vegetables

    b. More than two servings of fruits and vegetables

    c. At least two servings of fruits and vegetables

2. Find the probability that Rick does not eat exactly two servings of fruits and vegetables.

Example 2

Luis works in an office, and the phone rings occasionally. The possible number of phone calls he receives in an afternoon and their probabilities are given in the table below.

Number of Phone Calls	0	1	2	3	4
Probability	$\frac{1}{6}$	$\frac{1}{6}$	$\frac{2}{9}$	$\frac{1}{3}$	$\frac{1}{9}$

    a. Find the probability that Luis receives 3 or 4 phone calls.

EUREKA
MATH™

b. Find the probability that Luis receives fewer than 2 phone calls.

c. Find the probability that Luis receives 2 or fewer phone calls.

d. Find the probability that Luis does not receive 4 phone calls.

Exercises 3–7

When Jenna goes to the farmers' market, she also usually buys some broccoli. The possible number of heads of broccoli that she buys and the probabilities are given in the table below.

Number of Heads of Broccoli	0	1	2	3	4
Probability	$\frac{1}{12}$	$\frac{1}{6}$	$\frac{5}{12}$	$\frac{1}{4}$	$\frac{1}{12}$

3. Find the probability that Jenna:
   a. Buys exactly 3 heads of broccoli

   b. Does not buy exactly 3 heads of broccoli

   c. Buys more than 1 head of broccoli

   d. Buys at least 3 heads of broccoli

The diagram below shows a spinner designed like the face of a clock. The sectors of the spinner are colored red (R), blue (B), green (G), and yellow (Y).

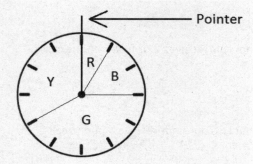

4. Writing your answers as fractions in lowest terms, find the probability that the pointer stops on the following colors.

   a.   Red:

   b.   Blue:

   c.   Green:

   d.   Yellow:

5. Complete the table of probabilities below.

Color	Red	Blue	Green	Yellow
Probability				

Lesson 5:   Chance Experiments with Outcomes That Are Not Equally Likely   **EUREKA MATH**™

6.   Find the probability that the pointer stops in either the blue region or the green region.

7.   Find the probability that the pointer does not stop in the green region.

Lesson Summary

In a probability experiment where the outcomes are not known to be equally likely, the formula for the probability of an event does not necessarily apply:

$$P(\text{event}) = \frac{\text{Number of outcomes in the event}}{\text{Number of outcomes in the sample space}}.$$

For example:

- To find the probability that the score is greater than 3, add the probabilities of all the scores that are greater than 3.
- To find the probability of not getting a score of 3, calculate $1 -$ (the probability of getting a 3).

Problem Set

1. The Gator Girls is a soccer team. The possible number of goals the Gator Girls will score in a game and their probabilities are shown in the table below.

Number of Goals	0	1	2	3	4
Probability	0.22	0.31	0.33	0.11	0.03

Find the probability that the Gator Girls:

a. Score more than two goals

b. Score at least two goals

c. Do not score exactly 3 goals

2. The diagram below shows a spinner. The pointer is spun, and the player is awarded a prize according to the color on which the pointer stops.

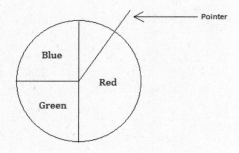

a. What is the probability that the pointer stops in the red region?

EUREKA MATH

b. Complete the table below showing the probabilities of the three possible results.

Color	Red	Green	Blue
Probability			

c. Find the probability that the pointer stops on green or blue.

d. Find the probability that the pointer does not stop on green.

3. Wayne asked every student in his class how many siblings (brothers and sisters) they had. The survey results are shown in the table below. (Wayne included himself in the results.)

Number of Siblings	0	1	2	3	4
Number of Students	4	5	14	6	3

(Note: The table tells us that 4 students had no siblings, 5 students had one sibling, 14 students had two siblings, and so on.)

a. How many students are there in Wayne's class, including Wayne?

b. What is the probability that a randomly selected student does not have any siblings? Write your answer as a fraction in lowest terms.

c. The table below shows the possible number of siblings and the probabilities of each number. Complete the table by writing the probabilities as fractions in lowest terms.

Number of Siblings	0	1	2	3	4
Probability					

d. Writing your answers as fractions in lowest terms, find the probability that the student:

i.   Has fewer than two siblings

ii. Has two or fewer siblings

iii. Does not have exactly one sibling

This page intentionally left blank

Lesson 6: Using Tree Diagrams to Represent a Sample Space and to Calculate Probabilities

Classwork

Suppose a girl attends a preschool where the students are studying primary colors. To help teach calendar skills, the teacher has each student maintain a calendar in his cubby. For each of the four days that the students are covering primary colors in class, students get to place a colored dot on their calendars: blue, yellow, or red. When the four days of the school week have passed (Monday–Thursday), what might the young girl's calendar look like?

One outcome would be four blue dots if the student chose blue each day. But consider that the first day (Monday) could be blue, and the next day (Tuesday) could be yellow, and Wednesday could be blue, and Thursday could be red. Or maybe Monday and Tuesday could be yellow, Wednesday could be blue, and Thursday could be red. Or maybe Monday, Tuesday, and Wednesday could be blue, and Thursday could be red, and so on and so forth.

As hard to follow as this seems now, we have only mentioned 3 of the 81 possible outcomes in terms of the four days of colors! Listing the other 78 outcomes would take several pages! Rather than listing outcomes in the manner described above (particularly when the situation has multiple stages, such as the multiple days in the case above), we often use a *tree diagram* to display all possible outcomes visually. Additionally, when the outcomes of each stage are the result of a chance experiment, tree diagrams are helpful for computing probabilities.

Example 1: Two Nights of Games

Imagine that a family decides to play a game each night. They all agree to use a tetrahedral die (i.e., a four-sided pyramidal die where each of four possible outcomes is equally likely—see the image at the end of this lesson) each night to randomly determine if they will play a board game (B) or a card game (C). The tree diagram mapping the possible overall outcomes over two consecutive nights will be developed below.

To make a tree diagram, first present all possibilities for the first stage (in this case, Monday).

Monday   *Tuesday*   *Outcome*

B

C

Then, from *each* branch of the first stage, attach all possibilities for the second stage (Tuesday).

<table>
<tr><td>**Monday**</td><td>**Tuesday**</td><td>**Outcome**</td></tr>
</table>

	B	BB
B		
	C	BC

	B	CB
C		
	C	CC

Note: If the situation has more than two stages, this process would be repeated until all stages have been presented.

a. If BB represents two straight nights of board games, what does CB represent?

b. List the outcomes where exactly one board game is played over two days. How many outcomes were there?

Lesson 6:   Using Tree Diagrams to Represent a Sample Space and to Calculate
Probabilities

EUREKA
MATH

Example 2: Two Nights of Games (with Probabilities)

In Example 1, each night's outcome is the result of a chance experiment (rolling the tetrahedral die). Thus, there is a probability associated with each night's outcome.

By multiplying the probabilities of the outcomes from each stage, we can obtain the probability for each "branch of the tree." In this case, we can figure out the probability of each of our four outcomes: BB, BC, CB, and CC.

For this family, a card game will be played if the die lands showing a value of 1, and a board game will be played if the die lands showing a value of 2, 3, or 4. This makes the probability of a board game (B) on a given night 0.75.

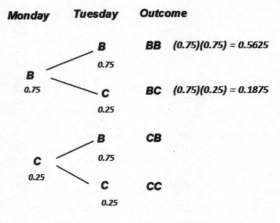

a.   The probabilities for two of the four outcomes are shown. Now, compute the probabilities for the two remaining outcomes.

b.   What is the probability that there will be exactly one night of board games over the two nights?

Exercises 1–3: Two Children

Two friends meet at a grocery store and remark that a neighboring family just welcomed their second child. It turns out that both children in this family are girls, and they are not twins. One of the friends is curious about what the chances are of having 2 girls in a family's first 2 births. Suppose that for each birth, the probability of a boy birth is 0.5 and the probability of a girl birth is also 0.5.

1. Draw a tree diagram demonstrating the four possible birth outcomes for a family with 2 children (no twins). Use the symbol B for the outcome of *boy* and G for the outcome of *girl*. Consider the first birth to be the first stage. (Refer to Example 1 if you need help getting started.)

2. Write in the probabilities of each stage's outcome to the tree diagram you developed above, and determine the probabilities for each of the 4 possible birth outcomes for a family with 2 children (no twins).

3. What is the probability of a family having 2 girls in this situation? Is that greater than or less than the probability of having exactly 1 girl in 2 births?

Lesson Summary

Tree diagrams can be used to organize outcomes in the sample space for chance experiments that can be thought of as being performed in multiple stages. Tree diagrams are also useful for computing probabilities of events with more than one outcome.

Problem Set

1. Imagine that a family of three (Alice, Bill, and Chester) plays bingo at home every night. Each night, the chance that any one of the three players will win is $\frac{1}{3}$.

   a. Using A for Alice wins, B for Bill wins, and C for Chester wins, develop a tree diagram that shows the nine possible outcomes for two consecutive nights of play.

   b. Is the probability that "Bill wins both nights" the same as the probability that "Alice wins the first night and Chester wins the second night"? Explain.

2. According to the Washington, D.C. Lottery's website for its Cherry Blossom Doubler instant scratch game, the chance of winning a prize on a given ticket is about 17%. Imagine that a person stops at a convenience store on the way home from work every Monday and Tuesday to buy a scratcher ticket to play the game.

   (Source: http://dclottery.com/games/scratchers/1223/cherry-blossom-doubler.aspx, accessed May 27, 2013)

   a. Develop a tree diagram showing the four possible outcomes of playing over these two days. Call stage 1 "Monday," and use the symbols W for a winning ticket and L for a non-winning ticket.

   b. What is the chance that the player will not win on Monday but will win on Tuesday?

   c. What is the chance that the player will win at least once during the two-day period?

Image of Tetrahedral Die

Source: http://commons.wikimedia.org/wiki/File:4-sided_dice_250.jpg

Photo by Fantasy, via Wikimedia Commons, is licensed under CC BY-SA 3.0, http://creativecommons.org/licenses/by-sa/3.0/deed.en.

This page intentionally left blank

Lesson 7: Calculating Probabilities of Compound Events

Classwork

A previous lesson introduced *tree diagrams* as an effective method of displaying the possible outcomes of certain multistage chance experiments. Additionally, in such situations, tree diagrams were shown to be helpful for computing probabilities.

In those previous examples, diagrams primarily focused on cases with two stages. However, the basic principles of tree diagrams can apply to situations with more than two stages.

Example 1: Three Nights of Games

Recall a previous example where a family decides to play a game each night, and they all agree to use a tetrahedral die (a four-sided die in the shape of a pyramid where each of four possible outcomes is equally likely) each night to randomly determine if the game will be a board (B) or a card (C) game. The tree diagram mapping the possible overall outcomes over two consecutive nights was as follows:

Monday	**Tuesday**	**Outcome**

```
                    B          BB

          B

                    C          BC


                    B          CB

          C

                    C          CC
```

But how would the diagram change if you were interested in mapping the possible overall outcomes over three consecutive nights? To accommodate this additional third stage, you would take steps similar to what you did before. You would attach all possibilities for the third stage (Wednesday) to each branch of the previous stage (Tuesday).

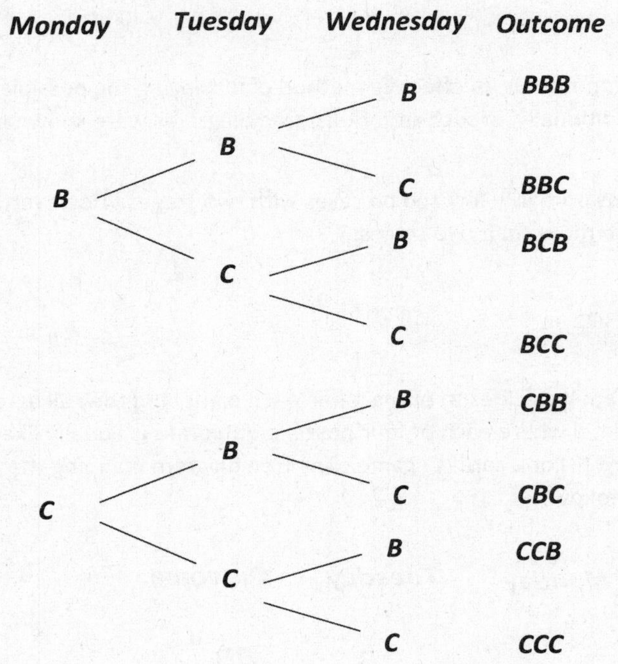

Monday	Tuesday	Wednesday	Outcome

Exercises 1–3

1. If BBB represents three straight nights of board games, what does CBB represent?

2. List all outcomes where exactly two board games were played over three days. How many outcomes were there?

3. There are eight possible outcomes representing the three nights. Are the eight outcomes representing the three nights equally likely? Why or why not?

EUREKA
MATH

©2015 Great Minds. eureka-math.org
G7-M5-SE-B3-1.3.1-01.2016

Example 2: Three Nights of Games (with Probabilities)

In Example 1, each night's outcome is the result of a chance experiment (rolling the four-sided die). Thus, there is a probability associated with each night's outcome.

By multiplying the probabilities of the outcomes from each stage, you can obtain the probability for each "branch of the tree." In this case, you can figure out the probability of each of our eight outcomes.

For this family, a card game will be played if the die lands showing a value of 1, and a board game will be played if the die lands showing a value of 2, 3, or 4. This makes the probability of a board game (B) on a given night 0.75.

Let's use a tree to examine the probabilities of the outcomes for the three days.

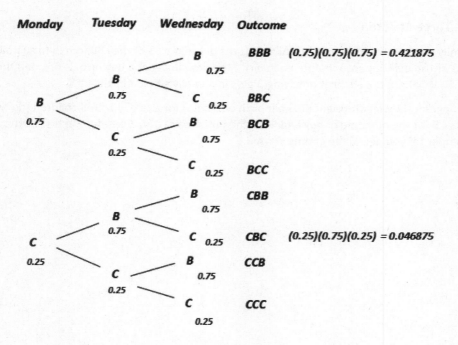

Exercises 4–6

4. Probabilities for two of the eight outcomes are shown. Calculate the approximate probabilities for the remaining six outcomes.

5. What is the probability that there will be exactly two nights of board games over the three nights?

6. What is the probability that the family will play at least one night of card games?

Exercises 7–10: Three Children

A neighboring family just welcomed their third child. It turns out that all 3 of the children in this family are girls, and they are not twins or triplets. Suppose that for each birth, the probability of a boy birth is 0.5, and the probability of a girl birth is also 0.5. What are the chances of having 3 girls in a family's first 3 births?

7. Draw a tree diagram showing the eight possible birth outcomes for a family with 3 children (no twins or triplets). Use the symbol B for the outcome of *boy* and G for the outcome of *girl*. Consider the first birth to be the first stage. (Refer to Example 1 if you need help getting started.)

8. Write in the probabilities of each stage's outcomes in the tree diagram you developed above, and determine the probabilities for each of the eight possible birth outcomes for a family with 3 children (no twins or triplets).

Lesson 5:   Calculating Probabilities of Compound Events

EUREKA MATH

9. What is the probability of a family having 3 girls in this situation? Is that greater than or less than the probability of having exactly 2 girls in 3 births?

10. What is the probability of a family of 3 children having at least 1 girl?

Lesson Summary

The use of tree diagrams is not limited to cases of just two stages. For more complicated experiments, tree diagrams are used to organize outcomes and to assign probabilities. The tree diagram is a visual representation of outcomes that involve more than one event.

Problem Set

1. According to the Washington, D.C. Lottery's website for its Cherry Blossom Double instant scratch game, the chance of winning a prize on a given ticket is about 17%. Imagine that a person stops at a convenience store on the way home from work every Monday, Tuesday, and Wednesday to buy a scratcher ticket and plays the game.

   (Source: http://dclottery.com/games/scratchers/1223/cherry-blossom-doubler.aspx, accessed May 27, 2013)

   a. Develop a tree diagram showing the eight possible outcomes of playing over these three days. Call stage one "Monday," and use the symbols W for a winning ticket and L for a non-winning ticket.

   b. What is the probability that the player will not win on Monday but will win on Tuesday and Wednesday?

   c. What is the probability that the player will win at least once during the 3-day period?

2. A survey company is interested in conducting a statewide poll prior to an upcoming election. They are only interested in talking to registered voters.

   Imagine that 55% of the registered voters in the state are male and 45% are female. Also, consider that the distribution of ages may be different for each group. In this state, 30% of male registered voters are age 18–24, 37% are age 25–64, and 33% are 65 or older. 32% of female registered voters are age 18–24, 26% are age 25–64, and 42% are 65 or older.

   The following tree diagram describes the distribution of registered voters. The probability of selecting a male registered voter age 18–24 is 0.165.

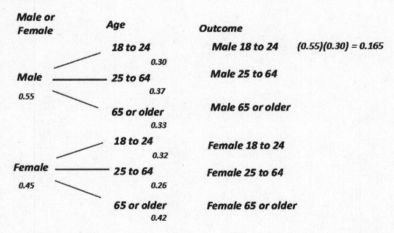

   a. What is the chance that the polling company will select a registered female voter age 65 or older?

   b. What is the chance that the polling company will select any registered voter age 18–24?

Lesson 8: The Difference Between Theoretical Probabilities and Estimated Probabilities

Classwork

Have you ever watched the beginning of a professional football game? After the traditional handshakes, a coin is tossed to determine which team gets to kick off first. The toss of a fair coin is often used to make decisions between two groups.

Example 1: Why a Coin?

Coins were discussed in previous lessons of this module. What is special about a coin? In most cases, a coin has two different sides: a head side (heads) and a tail side (tails). The sample space for tossing a coin is {heads, tails}. If each outcome has an equal chance of occurring when the coin is tossed, then the probability of getting heads is $\frac{1}{2}$, or 0.5. The probability of getting tails is also 0.5. Note that the sum of these probabilities is 1.

The probabilities formed using the sample space and what we know about coins are called the *theoretical* probabilities. Using observed relative frequencies is another method to estimate the probabilities of heads or tails. A relative frequency is the proportion derived from the number of the observed outcomes of an event divided by the total number of outcomes. Recall from earlier lessons that a relative frequency can be expressed as a fraction, a decimal, or a percent. Is the estimate of a probability from this method close to the theoretical probability? The following example investigates how relative frequencies can be used to estimate probabilities.

Beth tosses a coin 10 times and records her results. Here are the results from the 10 tosses:

Toss	1	2	3	4	5	6	7	8	9	10
Result	H	H	T	H	H	H	T	T	T	H

The total number of heads divided by the total number of tosses is the relative frequency of heads. It is the proportion of the time that heads occurred on these tosses. The total number of tails divided by the total number of tosses is the relative frequency of tails.

a. Beth started to complete the following table as a way to investigate the relative frequencies. For each outcome, the total number of tosses increased. The total number of heads or tails observed so far depends on the outcome of the current toss. Complete this table for the 10 tosses recorded in the previous table.

Toss	Outcome	Total Number of Heads So Far	Relative Frequency of Heads So Far (to the nearest hundredth)	Total Number of Tails So Far	Relative Frequency of Tails So Far (to the nearest hundredth)
1	H	1	$\frac{1}{1} = 1$	0	$\frac{0}{1} = 0$
2	H	2	$\frac{2}{2} = 1$	0	$\frac{0}{2} = 0$
3	T	2	$\frac{2}{3} \approx 0.67$	1	$\frac{1}{3} \approx 0.33$
4					
5					
6					
7					
8					
9					
10					

b. What is the sum of the relative frequency of heads and the relative frequency of tails for each row of the table?

Lesson 8:   The Difference Between Theoretical Probabilities and Estimated Probabilities

EUREKA MATH

c. Beth's results can also be displayed using a graph. Use the values of the relative frequency of heads so far from the table in part (a) to complete the graph below.

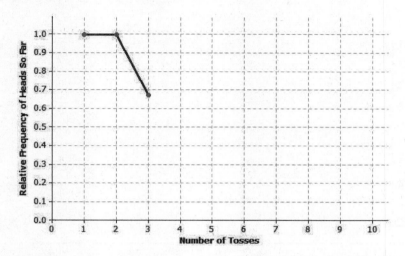

d. Beth continued tossing the coin and recording the results for a total of 40 tosses. Here are the results of the next 30 tosses:

Toss	11	12	13	14	15	16	17	18	19	20
Result	T	H	T	H	T	H	H	T	H	T

Toss	21	22	23	24	25	26	27	28	29	30
Result	H	T	T	H	T	T	T	T	H	T

Toss	31	32	33	34	35	36	37	38	39	40
Result	H	T	H	T	H	T	H	H	T	T

As the number of tosses increases, the relative frequency of heads changes. Complete the following table for the 40 coin tosses:

Number of Tosses	Total Number of Heads So Far	Relative Frequency of Heads So Far (to the nearest hundredth)
1		
5		
10		
15		
20		
25		
30		
35		
40		

e.   Use the relative frequency of heads so far from the table in part (d) to complete the graph below for the total number of tosses of 1, 5, 10, 15, 20, 25, 30, 35, and 40.

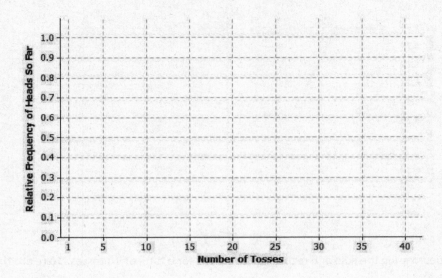

f.   What do you notice about the changes in the relative frequency of the number of heads so far as the number of tosses increases?

g.   If you tossed the coin 100 times, what do you think the relative frequency of heads would be? Explain your answer.

h.   Based on the graph and the relative frequencies, what would you estimate the probability of getting heads to be? Explain your answer.

Lesson 8:   The Difference Between Theoretical Probabilities and Estimated
              Probabilities

EUREKA
MATH™

i.   How close is your estimate in part (h) to the theoretical probability of 0.5? Would the estimate of this probability have been as good if Beth had only tossed the coin a few times instead of 40?

The value you gave in part (h) is an estimate of the theoretical probability and is called an *experimental* or *estimated* probability.

Exercises 1–8

Beth received nine more pennies. She securely taped them together to form a small stack. The top penny of her stack showed heads, and the bottom penny showed tails. If Beth tosses the stack, what outcomes could she observe?

1. Beth wanted to determine the probability of getting heads when she tosses the stack. Do you think this probability is the same as the probability of getting heads with just one coin? Explain your answer.

2. Make a sturdy stack of 10 pennies in which one end of the stack has a penny showing heads and the other end tails. Make sure the pennies are taped securely, or you may have a mess when you toss the stack. Toss the stack to observe possible outcomes. What is the sample space for tossing a stack of 10 pennies taped together? Do you think the probability of each outcome of the sample space is equal? Explain your answer.

3. Record the results of 10 tosses. Complete the following table of the relative frequencies of heads for your 10 tosses:

Toss	1	2	3	4	5	6	7	8	9	10
Result										
Relative Frequency of Heads So Far										

4. Based on the value of the relative frequencies of heads so far, what would you estimate the probability of getting heads to be?

5. Toss the stack of 10 pennies another 20 times. Complete the following table:

Toss	11	12	13	14	15	16	17	18	19	20
Result										

Toss	21	22	23	24	25	26	27	28	29	30
Result										

6. Summarize the relative frequency of heads so far by completing the following table:

Number of Tosses	Total Number of Heads So Far	Relative Frequency of Heads So Far (to the nearest hundredth)
1		
5		
10		
15		
20		
25		
30		

Lesson 8:   The Difference Between Theoretical Probabilities and Estimated Probabilities

7. Based on the relative frequencies for the 30 tosses, what is your estimate of the probability of getting heads? Can you compare this estimate to a theoretical probability like you did in the first example? Explain your answer.

8. Create another stack of pennies. Consider creating a stack using 5 pennies, 15 pennies, or 20 pennies taped together in the same way you taped the pennies to form a stack of 10 pennies. Again, make sure the pennies are taped securely, or you might have a mess!

    Toss the stack you made 30 times. Record the outcome for each toss:

Toss	1	2	3	4	5	6	7	8	9	10
Result										

Toss	11	12	13	14	15	16	17	18	19	20
Result										

Toss	21	22	23	24	25	26	27	28	29	30
Result										

Lesson Summary

- Observing the long-run relative frequency of an event from a chance experiment (or the proportion of an event derived from a long sequence of observations) approximates the theoretical probability of the event.

- After a long sequence of observations, the observed relative frequencies get close to the probability of the event occurring.

- When it is not possible to compute the theoretical probabilities of chance experiments, then the long-run relative frequencies (or the proportion of events derived from a long sequence of observations) can be used as estimated probabilities of events.

Problem Set

1. If you created a stack of 15 pennies taped together, do you think the probability of getting a heads on a toss of the stack would be different than for a stack of 10 pennies? Explain your answer.

2. If you created a stack of 20 pennies taped together, what do you think the probability of getting a heads on a toss of the stack would be? Explain your answer.

3. Based on your work in this lesson, complete the following table of the relative frequencies of heads for the stack you created:

Number of Tosses	Total Number of Heads So Far	Relative Frequency of Heads So Far (to the nearest hundredth)
1		
5		
10		
15		
20		
25		
30		

4. What is your estimate of the probability that your stack of pennies will land heads up when tossed? Explain your answer.

5. Is there a theoretical probability you could use to compare to the estimated probability? Explain your answer.

Lesson 9: Comparing Estimated Probabilities to Probabilities Predicted by a Model

Classwork

Exploratory Challenge: Game Show—Picking Blue!

Imagine, for a moment, the following situation: You and your classmates are contestants on a quiz show called *Picking Blue!* There are two bags in front of you, Bag A and Bag B. Each bag contains red and blue chips. You are told that one of the bags has exactly the same number of blue chips as red chips. But you are told nothing about the ratio of blue to red chips in the other bag.

Each student in your class will be asked to select either Bag A or Bag B. Starting with Bag A, a chip is randomly selected from the bag. If a blue chip is drawn, all of the students in your class who selected Bag A win a blue token. The chip is put back in the bag. After mixing up the chips in the bag, another chip is randomly selected from the bag. If the chip is blue, the students who picked Bag A win another blue token. After the chip is placed back into the bag, the process continues until a red chip is picked. When a red chip is picked, the game moves to Bag B. A chip from the Bag B is then randomly selected. If it is blue, all of the students who selected Bag B win a blue token. But if the chip is red, the game is over. Just like for Bag A, if the chip is blue, the process repeats until a red chip is picked from the bag. When the game is over, the students with the greatest number of blue tokens are considered the winning team.

Without any information about the bags, you would probably select a bag simply by guessing. But surprisingly, the show's producers are going to allow you to do some research before you select a bag. For the next 20 minutes, you can pull a chip from either one of the two bags, look at the chip, and then put the chip back in the bag. You can repeat this process as many times as you want within the 20 minutes. At the end of 20 minutes, you must make your final decision and select which of the bags you want to use in the game.

Getting Started

Assume that the producers of the show do not want to give away a lot of their blue tokens. As a result, if one bag has the same number of red and blue chips, do you think the other bag would have more or fewer blue chips than red chips? Explain your answer.

Planning the Research

Your teacher will provide you with two bags labeled A and B. You have 20 minutes to experiment with pulling chips one at a time from the bags. After you examine a chip, you must put it back in the bag. Remember, no peeking in the bags, as that will disqualify you from the game. You can pick chips from just one bag, or you can pick chips from one bag and then the other bag.

Use the results from 20 minutes of research to determine which bag you will choose for the game.

Provide a description outlining how you will carry out your research.

Carrying Out the Research

Share your plan with your teacher. Your teacher will verify whether your plan is within the rules of the quiz show. Approving your plan does not mean, however, that your teacher is indicating that your research method offers the most accurate way to determine which bag to select. If your teacher approves your research, carry out your plan as outlined. Record the results from your research, as directed by your teacher.

Playing the Game

After the research has been conducted, the competition begins. First, your teacher will shake up Bag A. A chip is selected. If the chip is blue, all students who selected Bag A win an imaginary blue token. The chip is put back in the bag, and the process continues. When a red chip is picked from Bag A, students selecting Bag A have completed the competition. Your teacher will now shake up Bag B. A chip is selected. If it is blue, all students who selected Bag B win an imaginary blue token. The process continues until a red chip is picked. At that point, the game is over.

How many blue tokens did you win?

©2015 Great Minds. eureka-math.org
G7-M5-SE-B3-1.3.1-01.2016

Examining Your Results

At the end of the game, your teacher will open the bags and reveal how many blue and red chips were in each bag. Answer the questions that follow. After you have answered these questions, discuss them with your class.

1. Before you played the game, what were you trying to learn about the bags from your research?

2. What did you expect to happen when you pulled chips from the bag with the same number of blue and red chips? Did the bag that you thought had the same number of blue and red chips yield the results you expected?

3. How confident were you in predicting which bag had the same number of blue and red chips? Explain.

4. What bag did you select to use in the competition, and why?

5. If you were the show's producers, how would you make up the second bag? (Remember, one bag has the same number of red and blue chips.)

6. If you picked a chip from Bag B 100 times and found that you picked each color exactly 50 times, would you know for sure that Bag B was the one with equal numbers of each color?

Lesson Summary

- The long-run relative frequencies can be used as estimated probabilities of events.
- Collecting data on a game or chance experiment is one way to estimate the probability of an outcome.
- The more data collected on the outcomes from a game or chance experiment, the closer the estimates of the probabilities are likely to be the actual probabilities.

Problem Set

Jerry and Michael played a game similar to *Picking Blue!* The following results are from their research using the same two bags:

Jerry's research:

	Number of Red Chips Picked	Number of Blue Chips Picked
Bag A	2	8
Bag B	3	7

Michael's research:

	Number of Red Chips Picked	Number of Blue Chips Picked
Bag A	28	12
Bag B	22	18

1. If all you knew about the bags were the results of Jerry's research, which bag would you select for the game? Explain your answer.

2. If all you knew about the bags were the results of Michael's research, which bag would you select for the game? Explain your answer.

3. Does Jerry's research or Michael's research give you a better indication of the makeup of the blue and red chips in each bag? Explain why you selected this research.

4. Assume there are 12 chips in each bag. Use either Jerry's or Michael's research to estimate the number of red and blue chips in each bag. Then, explain how you made your estimates.

   Bag A

   Number of red chips:

   Number of blue chips:

   Bag B

   Number of red chips:

   Number of blue chips:

5. In a different game of *Picking Blue!*, two bags each contain red, blue, green, and yellow chips. One bag contains the same number of red, blue, green, and yellow chips. In the second bag, half of the chips are blue. Describe a plan for determining which bag has more blue chips than any of the other colors.

Lesson 10: Conducting a Simulation to Estimate the Probability of an Event

Classwork

In previous lessons, you estimated probabilities of events by collecting data empirically or by establishing a theoretical probability model. There are real problems for which those methods may be difficult or not practical to use. Simulation is a procedure that will allow you to answer questions about real problems by running experiments that closely resemble the real situation.

It is often important to know the probabilities of real-life events that may not have known theoretical probabilities. Scientists, engineers, and mathematicians design simulations to answer questions that involve topics such as diseases, water flow, climate changes, or functions of an engine. Results from the simulations are used to estimate probabilities that help researchers understand problems and provide possible solutions to these problems.

Example 1: Families

How likely is it that a family with three children has all boys or all girls?

Let's assume that a child is equally likely to be a boy or a girl. Instead of observing the result of actual births, a toss of a fair coin could be used to simulate a birth. If the toss results in heads (H), then we could say a boy was born; if the toss results in tails (T), then we could say a girl was born. If the coin is fair (i.e., heads and tails are equally likely), then getting a boy or a girl is equally likely.

Exercises 1–2

Suppose that a family has three children. To simulate the genders of the three children, the coin or number cube or a card would need to be used three times, once for each child. For example, three tosses of the coin resulted in HHT, representing a family with two boys and one girl. Note that HTH and THH also represent two boys and one girl.

1. Suppose that when a prime number (P) is rolled on the number cube, it simulates a boy birth, and a non-prime (N) simulates a girl birth. Using such a number cube, list the outcomes that would simulate a boy birth and those that simulate a girl birth. Are the boy and girl birth outcomes equally likely?

2. Suppose that one card is drawn from a regular deck of cards. A red card (R) simulates a boy birth, and a black card (B) simulates a girl birth. Describe how a family of three children could be simulated.

Example 2

Simulation provides an estimate for the probability that a family of three children would have three boys or three girls by performing three tosses of a fair coin many times. Each sequence of three tosses is called a *trial*. If a trial results in either HHH or TTT, then the trial represents all boys or all girls, which is the event that we are interested in. These trials would be called a *success*. If a trial results in any other order of H's and T's, then it is called a *failure*.

The estimate for the probability that a family has either three boys or three girls based on the simulation is the number of successes divided by the number of trials. Suppose 100 trials are performed, and that in those 100 trials, 28 resulted in either HHH or TTT. Then, the estimated probability that a family of three children has either three boys or three girls would be $\frac{28}{100}$, or 0.28.

Exercises 3–5

3. Find an estimate of the probability that a family with three children will have exactly one girl using the following outcomes of 50 trials of tossing a fair coin three times per trial. Use H to represent a boy birth and T to represent a girl birth.

HHT	HTH	HHH	TTH	THT	THT	HTT	HHH	TTH	HHH
HHT	TTT	HHT	TTH	HHH	HTH	THH	TTT	THT	THT
THT	HHH	THH	HTT	HTH	TTT	HTT	HHH	TTH	THT
THH	HHT	TTT	TTH	HTT	THH	HTT	HTH	TTT	HHH
HTH	HTH	THT	TTH	TTT	HHT	HHT	THT	TTT	HTT

EUREKA
MATH™

4. Perform a simulation of 50 trials by rolling a fair number cube in order to find an estimate of the probability that a family with three children will have exactly one girl.

   a. Specify what outcomes of one roll of a fair number cube will represent a boy and what outcomes will represent a girl.

   b. Simulate 50 trials, keeping in mind that one trial requires three rolls of the number cube. List the results of your 50 trials.

   c. Calculate the estimated probability.

5. Calculate the theoretical probability that a family with three children will have exactly one girl.

   a. List the possible outcomes for a family with three children. For example, one possible outcome is BBB (all three children are boys).

   b. Assume that having a boy and having a girl are equally likely. Calculate the theoretical probability that a family with three children will have exactly one girl.

c.   Compare it to the estimated probabilities found in parts (a) and (b).

Example 3: Basketball Player

Suppose that, on average, a basketball player makes about three out of every four foul shots. In other words, she has a 75% chance of making each foul shot she takes. Since a coin toss produces equally likely outcomes, it could not be used in a simulation for this problem.

Instead, a number cube could be used by specifying that the numbers 1, 2, or 3 represent a hit, the number 4 represents a miss, and the numbers 5 and 6 would be ignored. Based on the following 50 trials of rolling a fair number cube, find an estimate of the probability that she makes five or six of the six foul shots she takes.

441323	342124	442123	422313	441243
124144	333434	243122	232323	224341
121411	321341	111422	114232	414411
344221	222442	343123	122111	322131
131224	213344	321241	311214	241131
143143	243224	323443	324243	214322
214411	423221	311423	142141	411312
343214	123131	242124	141132	343122
121142	321442	121423	443431	214433
331113	311313	211411	433434	323314

Problem Set

1. A mouse is placed at the start of the maze shown below. If it reaches station B, it is given a reward. At each point where the mouse has to decide which direction to go, assume that it is equally likely to go in either direction. At each decision point 1, 2, 3, it must decide whether to go left (L) or right (R). It cannot go backward.

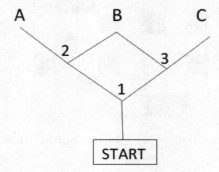

a. Create a theoretical model of probabilities for the mouse to arrive at terminal points A, B, and C.

i. List the possible paths of a sample space for the paths the mouse can take. For example, if the mouse goes left at decision point 1 and then right at decision point 2, then the path would be denoted LR.

ii. Are the paths in your sample space equally likely? Explain.

iii. What are the theoretical probabilities that a mouse reaches terminal points A, B, and C? Explain.

b. Based on the following set of simulated paths, estimate the probabilities that the mouse arrives at points A, B, and C.

RR	RR	RL	LL	LR	RL	LR	LL	LR	RR
LR	RL	LR	RR	RL	LR	RR	LL	RL	RL
LL	LR	LR	LL	RR	RR	RL	LL	RR	LR
RR	LR	RR	LR	LR	LL	LR	RL	RL	LL

c. How do the simulated probabilities in part (b) compare to the theoretical probabilities of part (a)?

2. Suppose that a dartboard is made up of the 8 × 8 grid of squares shown below. Also, suppose that when a dart is thrown, it is equally likely to land on any one of the 64 squares. A point is won if the dart lands on one of the 16 black squares. Zero points are earned if the dart lands in a white square.

a. For one throw of a dart, what is the probability of winning a point? Note that a point is won if the dart lands on a black square.

b. Lin wants to use a number cube to simulate the result of one dart. She suggests that 1 on the number cube could represent a win. Getting 2, 3, or 4 could represent no point scored. She says that she would ignore getting a 5 or 6. Is Lin's suggestion for a simulation appropriate? Explain why you would use it, or if not, how you would change it.

c. Suppose a game consists of throwing a dart three times. A trial consists of three rolls of the number cube. Based on Lin's suggestion in part (b) and the following simulated rolls, estimate the probability of scoring two points in three darts.

324	332	411	322	124
224	221	241	111	223
321	332	112	433	412
443	322	424	412	433
144	322	421	414	111
242	244	222	331	224
113	223	333	414	212
431	233	314	212	241
421	222	222	112	113
212	413	341	442	324

d. The theoretical probability model for winning 0, 1, 2, and 3 points in three throws of the dart as described in this problem is:

i.   Winning 0 points has a probability of 0.42.

ii. Winning 1 point has a probability of 0.42.

iii. Winning 2 points has a probability of 0.14.

iv. Winning 3 points has a probability of 0.02.

Use the simulated rolls in part (c) to build a model of winning 0, 1, 2, and 3 points, and compare it to the theoretical model.

Lesson 11: Conducting a Simulation to Estimate the Probability of an Event

Classwork

Example 1: Simulation

In the last lesson, we used coins, number cubes, and cards to carry out simulations. Another option is putting identical pieces of paper or colored disks into a container, mixing them thoroughly, and then choosing one.

For example, if a basketball player typically makes five out of eight foul shots, then a colored disk could be used to simulate a foul shot. A green disk could represent a made shot, and a red disk could represent a miss. You could put five green and three red disks in a container, mix them, and then choose one to represent a foul shot. If the color of the disk is green, then the shot is made. If the color of the disk is red, then the shot is missed. This procedure simulates one foul shot.

    a.   Using colored disks, describe how one at bat could be simulated for a baseball player who has a batting average of 0.300. Note that a batting average of 0.300 means the player gets a hit (on average) three times out of every ten times at bat. Be sure to state clearly what a color represents.

    b.   Using colored disks, describe how one at bat could be simulated for a player who has a batting average of 0.273. Note that a batting average of 0.273 means that on average, the player gets 273 hits out of 1,000 at bats.

Example 2: Using Random Number Tables

Why is using colored disks not practical for the situation described in Example 1(b)? Another way to carry out a simulation is to use a random number table, or a random number generator. In a random number table, the digits 0, 1, 2, 3, 4, 5, 6, 7, 8, and 9 occur equally often in the long run. Pages and pages of random numbers can be found online.

For example, here are three lines of random numbers. The space after every five digits is only for ease of reading. Ignore the spaces when using the table.

25256 65205 72597 00562 12683 90674 78923 96568 32177 33855

76635 92290 88864 72794 14333 79019 05943 77510 74051 87238

07895 86481 94036 12749 24005 80718 13144 66934 54730 77140

To use the random number table to simulate an at bat for the 0.273 hitter in Example 1(b), you could use a three-digit number to represent one at bat. The three-digit numbers 000–272 could represent a hit, and the three-digit numbers 273–999 could represent a non-hit. Using the random numbers above and starting at the beginning of the first line, the first three-digit random number is 252, which is between 000 and 272, so that simulated at bat is a hit. The next three-digit random number is 566, which is a non-hit.

Continuing on the first line of the random numbers above, what would the hit/non-hit outcomes be for the next six at bats? Be sure to state the random number and whether it simulates a hit or non-hit.

Example 3: Baseball Player

A batter typically gets to bat four times in a ball game. Consider the 0.273 hitter from the previous example. Use the following steps (and the random numbers shown above) to estimate that player's probability of getting at least three hits (three or four) in four times at bat.

a.   Describe what one trial is for this problem.

b.   Describe when a trial is called a success and when it is called a failure.

c.   Simulate 12 trials. (Continue to work as a class, or let students work with a partner.)

d.   Use the results of the simulation to estimate the probability that a 0.273 hitter gets three or four hits in four times at bat. Compare your estimate with other groups.

Example 4: Birth Month

In a group of more than 12 people, is it likely that at least two people, maybe more, will have the same birth month? Why? Try it in your class.

Now, suppose that the same question is asked for a group of only seven people. Are you likely to find some groups of seven people in which there is a match but other groups in which all seven people have different birth months? In the following exercises, you will estimate the probability that at least two people in a group of seven were born in the same month.

Exercises 1–4

1. What might be a good way to generate outcomes for the birth month problem—using coins, number cubes, cards, spinners, colored disks, or random numbers?

2. How would you simulate one trial of seven birth months?

3. How is a success determined for your simulation?

4. How is the simulated estimate determined for the probability that a least two in a group of seven people were born in the same month?

Problem Set

1. A model airplane has two engines. It can fly if one engine fails but is in serious trouble if both engines fail. The engines function independently of one another. On any given flight, the probability of a failure is 0.10 for each engine. Design a simulation to estimate the probability that the airplane will be in serious trouble the next time it goes up.

    a. How would you simulate the status of an engine?

    b. What constitutes a trial for this simulation?

    c. What constitutes a success for this simulation?

    d. Carry out 50 trials of your simulation, list your results, and calculate an estimate of the probability that the airplane will be in serious trouble the next time it goes up.

2. In an effort to increase sales, a cereal manufacturer created a really neat toy that has six parts to it. One part is put into each box of cereal. Which part is in a box is not known until the box is opened. You can play with the toy without having all six parts, but it is better to have the complete set. If you are really lucky, you might only need to buy six boxes to get a complete set. But if you are very unlucky, you might need to buy many, many boxes before obtaining all six parts.

    a. How would you represent the outcome of purchasing a box of cereal, keeping in mind that there are six different parts? There is one part in each box.

    b. If it was stated that a customer would have to buy at least 10 boxes of cereal to collect all six parts, what constitutes a trial in this problem?

    c. What constitutes a success in a trial in this problem?

    d. Carry out 15 trials, list your results, and compute an estimate of the probability that it takes the purchase of 10 or more boxes to get all six parts.

3. Suppose that a type A blood donor is needed for a certain surgery. Carry out a simulation to answer the following question: If 40% of donors have type A blood, what is an estimate of the probability that it will take at least four donors to find one with type A blood?

   a. How would you simulate a blood donor having or not having type A?

   b. What constitutes a trial for this simulation?

   c. What constitutes a success for this simulation?

   d. Carry out 15 trials, list your results, and compute an estimate for the probability that it takes at least four donors to find one with type A blood.

EUREKA
MATH™

Lesson 12: Applying Probability to Make Informed Decisions

Classwork

Example 1: Number Cube

Your teacher gives you a number cube with numbers 1–6 on its faces. You have never seen that particular cube before. You are asked to state a theoretical probability model for rolling it once. A probability model consists of the list of possible outcomes (the sample space) and the theoretical probabilities associated with each of the outcomes. You say that the probability model might assign a probability of $\frac{1}{6}$ to each of the possible outcomes, but because you have never seen this particular cube before, you would like to roll it a few times. (Maybe it is a trick cube.) Suppose your teacher allows you to roll it 500 times, and you get the following results:

Outcome	1	2	3	4	5	6
Frequency	77	92	75	90	76	90

Exercises 1–2

1. If the equally likely model was correct, about how many of each outcome would you expect to see if the cube is rolled 500 times?

2. Based on the data from the 500 rolls, how often were odd numbers observed? How often were even numbers observed?

Example 2: Probability Model

Two black balls and two white balls are put in a small cup whose bottom allows the four balls to fit snugly. After shaking the cup well, two patterns of colors are possible, as shown. The pattern on the left shows the similar colors are opposite each other, and the pattern on the right shows the similar colors are next to or adjacent to each other.

Philippe is asked to specify a probability model for the chance experiment of shaking the cup and observing the pattern. He thinks that because there are two outcomes—like heads and tails on a coin—that the outcomes should be equally likely. Sylvia isn't so sure that the equally likely model is correct, so she would like to collect some data before deciding on a model.

Exercise 3

3. Collect data for Sylvia. Carry out the experiment of shaking a cup that contains four balls, two black and two white, observing, and recording whether the pattern is opposite or adjacent. Repeat this process 20 times. Then, combine the data with those collected by your classmates.

    Do your results agree with Philippe's equally likely model, or do they indicate that Sylvia had the right idea? Explain.

Exercises 4–5

There are three popular brands of mixed nuts. Your teacher loves cashews, and in his experience of having purchased these brands, he suggests that not all brands have the same percentage of cashews. One has around 20% cashews, one has 25%, and one has 35%.

Your teacher has bags labeled A, B, and C representing the three brands. The bags contain red beads representing cashews and brown beads representing other types of nuts. One bag contains 20% red beads, another 25% red beads, and the third has 35% red beads. You are to determine which bag contains which percentage of cashews. You cannot just open the bags and count the beads.

4. Work as a class to design a simulation. You need to agree on what an outcome is, what a trial is, what a success is, and how to calculate the estimated probability of getting a cashew. Base your estimate on 50 trials.

5. Your teacher will give your group one of the bags labeled A, B, or C. Using your plan from part (a), collect your data. Do you think you have the 20%, 25%, or 35% cashews bag? Explain.

Exercises 6–8

Suppose you have two bags, A and B, in which there are an equal number of slips of paper. Positive numbers are written on the slips. The numbers are not known, but they are whole numbers between 1 and 75, inclusive. The same number may occur on more than one slip of paper in a bag.

These bags are used to play a game. In this game, you choose one of the bags and then choose one slip from that bag. If you choose Bag A and the number you choose from it is a prime number, then you win. If you choose Bag B and the number you choose from it is a power of 2, you win. Which bag should you choose?

6. Emma suggests that it does not matter which bag you choose because you do not know anything about what numbers are inside the bags. So, she thinks that you are equally likely to win with either bag. Do you agree with her? Explain.

7. Aamir suggests that he would like to collect some data from both bags before making a decision about whether or not the model is equally likely. Help Aamir by drawing 50 slips from each bag, being sure to replace each one before choosing again. Each time you draw a slip, record whether it would have been a winner or not. Using the results, what is your estimate for the probability of drawing a prime number from Bag A and drawing a power of 2 from Bag B?

8. If you were to play this game, which bag would you choose? Explain why you would pick this bag.

EUREKA MATH

Problem Set

1. Some M&M's® are "defective." For example, a defective M&M® may have its *M* missing, or it may be cracked, broken, or oddly shaped. Is the probability of getting a defective M&M® higher for peanut M&M's® than for plain M&M's®?

    Gloriann suggests the probability of getting a defective plain M&M® is the same as the probability of getting a defective peanut M&M®. Suzanne does not think this is correct because a peanut M&M® is bigger than a plain M&M®, and therefore has a greater opportunity to be damaged.

    a. Simulate inspecting a plain M&M® by rolling two number cubes. Let a sum of 7 or 11 represent a defective plain M&M® and the other possible rolls represent a plain M&M® that is not defective. Do 50 trials, and compute an estimate of the probability that a plain M&M® is defective. Record the 50 outcomes you observed. Explain your process.

    b. Simulate inspecting a peanut M&M® by selecting a card from a well-shuffled deck of cards. Let a one-eyed face card and clubs represent a defective peanut M&M® and the other cards represent a peanut M&M® that is not defective. Be sure to replace the chosen card after each trial and to shuffle the deck well before choosing the next card. Note that the one-eyed face cards are the king of diamonds, jack of hearts, and jack of spades. Do 20 trials, and compute an estimate of the probability that a peanut M&M® is defective. Record the list of 20 cards that you observed. Explain your process.

    c. For this problem, suppose that the two simulations provide accurate estimates of the probability of a defective M&M® for plain and peanut M&M's®. Compare your two probability estimates, and decide whether Gloriann's belief is reasonable that the defective probability is the same for both types of M&M's®. Explain your reasoning.

2. One at a time, mice are placed at the start of the maze shown below. There are four terminal stations at A, B, C, and D. At each point where a mouse has to decide in which direction to go, assume that it is equally likely for it to choose any of the possible directions. A mouse cannot go backward.

In the following simulated trials, L stands for left, R for right, and S for straight. Estimate the probability that a mouse finds station C where the food is. No food is at A, B, or D. The following data were collected on 50 simulated paths that the mice took.

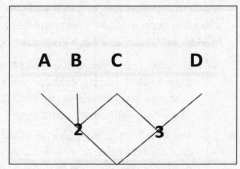

    LR   RL   RL   LL   LS   LS   RL   RR   RR   RL

    RL   LR   LR   RR   LR   LR   LL   LS   RL   LR

    RR   LS   RL   RR   RL   LR   LR   LL   LS   RR

    RL   RL   RL   RR   RR   RR   LR   LL   LL   RR

    RR   LS   RR   LR   RR   RR   LL   RR   LS   LS

a. What paths constitute a success, and what paths constitute a failure?

b. Use the data to estimate the probability that a mouse finds food. Show your calculation.

c. Paige suggests that it is equally likely that a mouse gets to any of the four terminal stations. What does your simulation suggest about whether her equally likely model is believable? If it is not believable, what do your data suggest is a more believable model?

d. Does your simulation support the following theoretical probability model? Explain.

    i.   The probability a mouse finds terminal point A is 0.167.

    ii. The probability a mouse finds terminal point B is 0.167.

    iii. The probability a mouse finds terminal point C is 0.417.

    iv. The probability a mouse finds terminal point D is 0.250.

EUREKA MATH™

Lesson 13: Populations, Samples, and Generalizing from a Sample to a Population

Classwork

In this lesson, you will learn about collecting data from a sample that is selected from a population. You will also learn about summary values for both a population and a sample and think about what can be learned about the population by looking at a sample from that population.

Exercises 1–4: Collecting Data

1. Describe what you would do if you had to collect data to investigate the following statistical questions using either a sample statistic or a population characteristic. Explain your reasoning in each case.

   a. How might you collect data to answer the question, "Does the soup taste good?"

   b. How might you collect data to answer the question, "How many movies do students in your class see in a month?"

   c. How might you collect data to answer the question, "What is the median price of a home in our town?"

   d. How might you collect data to answer the question, "How many pets do people own in my neighborhood?"

   e. How might you collect data to answer the question, "What is the typical number of absences in math classes at your school on a given day?"

f. How might you collect data to answer the question, "What is the typical life span of a particular brand of flashlight battery?"

g. How might you collect data to answer the question, "What percentage of girls and of boys in your school have a curfew?"

h. How might you collect data to answer the question, "What is the most common blood type of students in my class?"

A *population* is the entire set of objects (e.g., people, animals, and plants) from which data might be collected. A *sample* is a subset of the population. Numerical summary values calculated using data from an entire population are called *population characteristics*. Numerical summary values calculated using data from a sample are called *statistics*.

2. For which of the scenarios in Exercise 1 did you describe collecting data from a population and which from a sample?

3. Think about collecting data in the scenarios above. Give at least two reasons you might want to collect data from a sample rather than from the entire population.

4. Make up a result you might get in response to the situations in Exercise 1, and identify whether the result would be based on a population characteristic or a sample statistic.

   a. Does the soup taste good?

   b. How many movies do students in your class see in a month?

   c. What is the median price of a home in our town?

   d. How many pets do people own in my neighborhood?

   e. What is the typical number of absences in math classes at your school on a given day?

   f. What is the typical life span of a particular brand of flashlight battery?

   g. What percentage of girls and of boys in your school have a curfew?

   h. What is the most common blood type of students in my class?

Exercise 5: Population or Sample?

5.   Indicate whether the following statements are summarizing data collected to answer a statistical question from a population or from a sample. Identify references in the statement as population characteristics or sample statistics.

   a.   54% of the responders to a poll at a university indicated that wealth needed to be distributed more evenly among people.

   b.   Are students in the Bay Shore School District proficient on the state assessments in mathematics? In 2013, after all the tests taken by the students in the Bay Shore schools were evaluated, over 52% of those students were at or above proficient on the state assessment.

   c.   Does talking on mobile phones while driving distract people? Researchers measured the reaction times of 38 study participants as they talked on mobile phones and found that the average level of distraction from their driving was rated 2.25 out of 5.

   d.   Did most people living in New York in 2010 have at least a high school education? Based on the data collected from all New York residents in 2010 by the U.S. Census Bureau, 84.6% of people living in New York had at least a high school education.

   e.   Were there more deaths than births in the United States between July 2011 and July 2012? Data from a health service agency indicated that there were 2% more deaths than births in the United States during that time frame.

f.   What is the fifth best-selling book in the United States? Based on the sales of books in the United States, the fifth best-selling book was *Oh, the Places You'll Go!* by Dr. Seuss.

Exercises 6–8: A Census

6.   When data are collected from an entire population, it is called a *census*. The United States takes a census of its population every ten years, with the most recent one occurring in 2010. Go to http://www.census.gov to find the history of the U.S. census.

a.   Identify three things that you found to be interesting.

b.   Why is the census important in the United States?

7.   Go to the site: www.census.gov/2010census/popmap/ipmtext.php?fl=36.
Select the state of New York.

a.   How many people were living in New York for the 2010 census?

b.   Estimate the ratio of those 65 and older to those under 18 years old. Why is this important to think about?

c.   Is the ratio a population characteristic or a statistic? Explain your thinking.

8. The American Community Survey (ACS) takes samples from a small percentage of the U.S. population in years between the censuses. (www.census.gov/acs/www/about_the_survey/american_community_survey/)

   a. What is the difference between the way the ACS collects information about the U.S. population and the way the U.S. Census Bureau collects information?

   b. In 2011, the ACS sampled workers living in New York about commuting to work each day. Why do you think these data are important for the state to know?

   c. Suppose that from a sample of 200,000 New York workers, 32,400 reported traveling more than an hour to work each day. From this information, statisticians determined that between 16% and 16.4% of the workers in the state traveled more than an hour to work every day in 2011. If there were 8,437,512 workers in the entire population, about how many traveled more than an hour to work each day?

   d. Reasoning from a sample to the population is called *making an inference* about a population characteristic. Identify the statistic involved in making the inference in part (c).

   e. The data about traveling time to work suggest that across the United States typically between 79.8% and 80% of commuters travel alone, 10% to 10.2% carpool, and 4.9% to 5.1% use public transportation. Survey your classmates to find out how a worker in their families gets to work. How do the results compare to the national data? What might explain any differences?

EUREKA
MATH™

Lesson Summary

When data from a population are used to calculate a numerical summary, the value is called a *population characteristic*. When data from a sample are used to calculate a numerical summary, the value is called a *sample statistic*. Sample statistics can be used to learn about population characteristics.

Problem Set

1. The lunch program at Blake Middle School is being revised to align with the new nutritional standards that reduce calories and increase servings of fruits and vegetables. The administration decided to do a census of all students at Blake Middle School by giving a survey to all students about the school lunches.

   http://frac.org/federal-foodnutrition-programs/school-breakfast-program/school-meal-nutrition-standards

   a. Name some questions that you would include in the survey. Explain why you think those questions would be important to ask.

   b. Read through the paragraph below that describes some of the survey results. Then, identify the population characteristics and the sample statistics.

   > About $\frac{3}{4}$ of the students surveyed eat the school lunch regularly. The median number of days per month that students at Blake Middle School ate a school lunch was 18 days. 36% of students responded that their favorite fruit is bananas. The survey results for Tanya's seventh-grade homeroom showed that the median number of days per month that her classmates ate lunch at school was 22, and only 20% liked bananas. The fiesta salad was approved by 78% of the group of students who tried it, but when it was put on the lunch menu, only 40% of the students liked it. Of the seventh graders as a whole, 73% liked spicy jicama strips, but only 2 out of 5 of all the middle school students liked them.

2. For each of the following questions, (1) describe how you would collect data to answer the question, and (2) describe whether it would result in a sample statistic or a population characteristic.

   a. Where should the eighth-grade class go for its class trip?

   b. What is the average number of pets per family for families that live in your town?

   c. If people tried a new diet, what percentage would have an improvement in cholesterol reading?

   d. What is the average grade point of students who got accepted to a particular state university?

   e. What is a typical number of home runs hit in a particular season for major league baseball players?

3. Identify a question that would lead to collecting data from the given set as a population and a question where the data could be a sample from a larger population.

   a. All students in your school

   b. Your state

Lesson 13:   Populations, Samples, and Generalizing from a Sample to a Population

S.89

4. Suppose that researchers sampled attendees of a certain movie and found that the mean age was 17 years old. Based on this observation, which of the following would be most likely?

    a. The mean age of all of the people who went to see the movie was 17 years old.

    b. About a fourth of the people who went to see the movie were older than 51.

    c. The mean age of all people who went to see the movie would probably be in an interval around 17 years of age, that is, between 15 and 19.

    d. The median age of those who attended the movie was 17 years old as well.

5. The headlines proclaimed: "Education Impacts Work-Life Earnings Five Times More Than Other Demographic Factors, Census Bureau Reports." According to a U.S. Census Bureau study, education levels had more effect on earnings over a 40-year span in the workforce than any other demographic factor. www.census.gov/newsroom/releases/archives/education/cb11-153.html

    a. The article stated that the estimated impact on annual earnings between a professional degree and an eighth-grade education was roughly five times the impact of gender, which was $13,000. What would the difference in annual earnings be with a professional degree and with an eighth-grade education?

    b. Explain whether you think the data are from a population or a sample, and identify either the population characteristic or the sample statistic.

The History of the U.S. Census

The word *census* is Latin in origin and means to tax. The first U.S. census took place over 200 years ago, but the United States is certainly not the first country to implement a census. Based on archaeological records, it appears that the ancient Egyptians conducted a census as early as 3000 B.C.E.

The U.S. census is mandated by the U.S. Constitution in Article I, Section 2, which states, in part, "Representatives and direct Taxes shall be apportioned among the several States ... according to their respective Numbers .... The Number of Representatives shall not exceed one for every thirty thousand, but each State shall have at Least one Representative ...." The Constitution then specifies how to calculate the number of people in each state and how often the census should take place.

The U.S. census has been conducted every ten years since 1790, but as time has passed, our census has evolved. Not only have the types of questions changed but also the manner in which the data are collected and tabulated. Originally, the census had only a few questions, the purpose of which was to discern the number of people in each household and their ages. Presumably, these data were used to determine the number of men in each state who were available to go to war. Federal marshals were charged with the task of conducting this first census. After collecting data from their respective jurisdictions, the marshals sent the data to President Washington.

As time has passed, more questions have been added to the U.S. census. Today, the census includes questions designed to collect data in various fields such as manufacturing, commerce, and transportation, to name a few. Data that were once manually tabulated are now processed by computers. Home visits by census officials were once the norm, but now the census is conducted primarily through the U.S. Postal Service. Each household in the United States receives in the mail a copy of the census questionnaire to be completed by its head of household who then mails it back to the Census Bureau. Home visits are paid only to those individuals who do not return the questionnaire by the specified deadline.

The census is an important part of our Constitution. Today, the census not only tells us the population of each state, thereby determining the number of representatives that each state will have in the House of Representatives, but it also provides the U.S. government with very useful data that paint a picture of the current state of our population and how it has changed over the decades.

"U.S. Census History," *essortment*, accessed November 4, 2014, http://www.essortment.com/census-history-20901.html.

This page intentionally left blank

Lesson 14: Selecting a Sample

Classwork

As you learned in Lesson 13, sampling is a central concept in statistics. Examining every element in a population is usually impossible. So, research and articles in the media typically refer to a "sample" from a population. In this lesson, you will begin to think about how to choose a sample.

Exercises 1–2: What Is Random?

1. Write down a sequence of heads/tails you think would typically occur if you tossed a coin 20 times. Compare your sequence to the ones written by some of your classmates. How are they alike? How are they different?

2. Working with a partner, toss a coin 20 times, and write down the sequence of heads and tails you get.

   a. Compare your results with your classmates'.

   b. How are your results from actually tossing the coin different from the sequences you and your classmates wrote down?

    c.   Toni claimed she could make up a set of numbers that would be random. What would you say to her?

Exercises 3–11: Length of Words in the Poem "Casey at the Bat"

3.   Suppose you wanted to learn about the lengths of the words in the poem "Casey at the Bat." You plan to select a sample of eight words from the poem and use these words to answer the following statistical question: On average, how long is a word in the poem? What is the population of interest here?

4.   Look at the poem "Casey at the Bat" by Ernest Thayer, and select eight words you think are representative of words in the poem. Record the number of letters in each word you selected. Find the mean number of letters in the words you chose.

5.   A random sample is a sample in which every possible sample of the same size has an equal chance of being chosen. Do you think the set of words you wrote down was random? Why or why not?

6.   Working with a partner, follow your teacher's instructions for randomly choosing eight words. Begin with the title of the poem, and count a hyphenated word as one word.

    a.   Record the eight words you randomly selected, and find the mean number of letters in those words.

b. Compare the mean of your random sample to the mean you found in Exercise 4. Explain how you found the mean for each sample.

7. As a class, compare the means from Exercise 4 and the means from Exercise 6. Your teacher will provide a chart to compare the means. Record your mean from Exercise 4 and your mean for Exercise 6 on this chart.

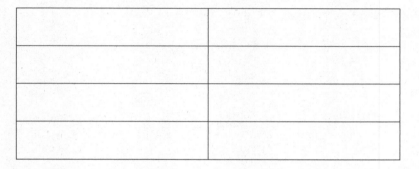

8. Do you think the means from Exercise 4 or the means from Exercise 6 are more representative of the mean of all of the words in the poem? Explain your choice.

9. The actual mean of the words in the poem "Casey at the Bat" is 4.2 letters. Based on the fact that the population mean is 4.2 letters, are the means from Exercise 4 or means from Exercise 6 a better representation of the mean of the population? Explain your answer.

10. How did the population mean of 4.2 letters compare to the mean of your random sample from Exercise 6 and to the mean you found in Exercise 4?

11. Summarize how you would estimate the mean number of letters in the words of another poem based on what you learned in the above exercises.

Problem Set

1. Would any of the following provide a random sample of letters used in the text of the book *Harry Potter and the Sorcerer's Stone* by J.K. Rowling? Explain your reasoning.

   a. Use the first letter of every word of a randomly chosen paragraph.

   b. Number all of the letters in the words in a paragraph of the book, cut out the numbers, and put them in a bag. Then, choose a random set of numbers from the bag to identify which letters you will use.

   c. Have a family member or friend write down a list of his favorite words, and count the number of times each of the letters occurs.

2. Indicate whether the following are random samples from the given population, and explain why or why not.

   a. Population: All students in school; the sample includes every fifth student in the hall outside of class.

   b. Population: Students in your class; the sample consists of students who have the letter *s* in their last names.

   c. Population: Students in your class; the sample is selected by putting their names in a hat and drawing the sample from the hat.

   d. Population: People in your neighborhood; the sample includes those outside in the neighborhood at 6:00 p.m.

   e. Population: Everyone in a room; the sample is selected by having everyone toss a coin, and those that result in heads are the sample.

3. Consider the two sample distributions of the number of letters in randomly selected words shown below:

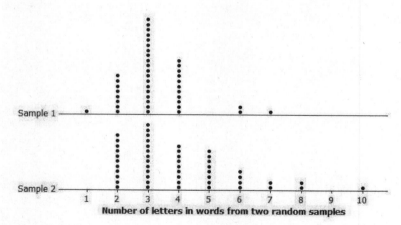

   a. Describe each distribution using statistical terms as much as possible.

   b. Do you think the two samples came from the same poem? Why or why not?

4. What questions about samples and populations might you want to ask if you saw the following headlines in a newspaper?

   a. "Peach Pop is the top flavor according to 8 out of 10 people."

   b. "Candidate X looks like a winner! 10 out of 12 people indicate they will vote for Candidate X."

   c. "Students overworked. Over half of 400 people surveyed think students spend too many hours on homework."

   d. "Action/adventure was selected as the favorite movie type by an overwhelming 75% of those surveyed."

This page intentionally left blank

Lesson 15: Random Sampling

Classwork

In this lesson, you will investigate taking random samples and how random samples from the same population vary.

Exercises 1–5: Sampling Pennies

1. Do you think different random samples from the same population will be fairly similar? Explain your reasoning.

2. The plot below shows the number of years since being minted (the penny age) for 150 pennies that JJ had collected over the past year. Describe the shape, center, and spread of the distribution.

Dot Plot of Population of Penny Ages

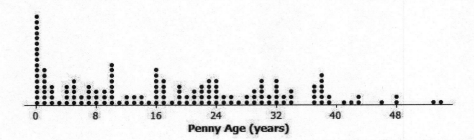

Penny Age (years)

3. Place ten dots on the number line that you think might be the distribution of a sample of 10 pennies from the jar.

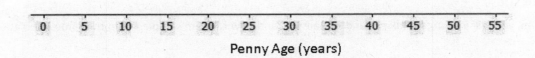

Penny Age (years)

4. Select a random sample of 10 pennies, and make a dot plot of the ages. Describe the distribution of the penny ages in your sample. How does it compare to the population distribution?

5. Compare your sample distribution to the sample distributions on the board.
    a. What do you observe?

    b. How does your sample distribution compare to those on the board?

Exercises 6–9: Grocery Prices and Rounding

6. Look over some of the grocery prices for this activity. Consider the following statistical question: "Do the store owners price the merchandise with cents that are closer to a higher dollar value or a lower dollar value?" Describe a plan that might answer that question that does not involve working with all 100 items.

7. Do the store owners price the merchandise with cents that are closer to a higher dollar value or a lower dollar value? To investigate this question in one situation, you will look at some grocery prices in weekly flyers and advertising for local grocery stores.

   a. How would you round $3.49 and $4.99 to the nearest dollar?

   b. If the advertised price was three for $4.35, how much would you expect to pay for one item?

   c. Do you think more grocery prices will round up or round down? Explain your thinking.

8. Follow your teacher's instructions to cut out the items and their prices from the weekly flyers and put them in a bag. Select a random sample of 25 items without replacement, and record the items and their prices in the table below.

Item	Price	Rounded	Item	Price	Rounded

Example of chart suggested:

Student	Number of Times the Prices Were Rounded to the Higher Value	Percent of Prices Rounded Up	Number of Times the Prices Were Rounded to the Lower Value
Bettina	20	80%	5

9. Round each of the prices in your sample to the nearest dollar, and count the number of times you rounded up and the number of times you rounded down.

   a. Given the results of your sample, how would you answer the question: Are grocery prices in the weekly ads at the local grocery closer to a higher dollar value or a lower dollar value?

   b. Share your results with classmates who used the same flyer or ads. Looking at the results of several different samples, how would you answer the question in part (a)?

   c. Identify the population, sample, and sample statistic used to answer the statistical question.

   d. Bettina says that over half of all the prices in the grocery store will round up. What would you say to her?

Problem Set

1. Look at the distribution of years since the pennies were minted from Example 1. Which of the following box plots seem like they might not have come from a random sample from that distribution? Explain your thinking.

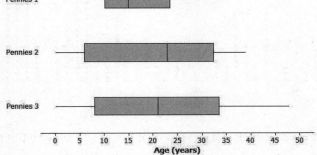

Box Plots of Three Random Samples of Penny Ages

2. Given the following sample of scores on a physical fitness test, from which of the following populations might the sample have been chosen? Explain your reasoning.

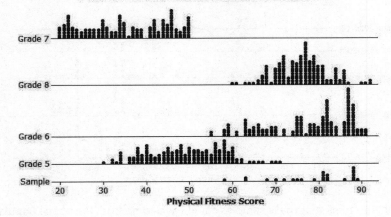

Dot Plots of Four Populations and One Sample

EUREKA
MATH™

3. Consider the distribution below:

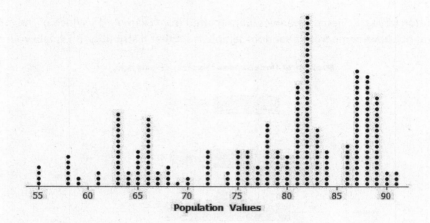

Population Values

a. What would you expect the distribution of a random sample of size 10 from this population to look like?

b. Random samples of different sizes that were selected from the population in part (a) are displayed below. How did your answer to part (a) compare to these samples of size 10?

Dot Plots of Five Samples of Different Sizes

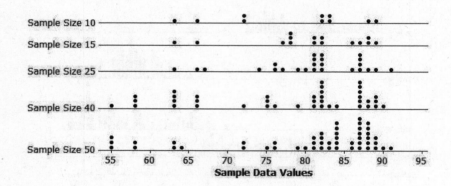

Sample Data Values

c. Why is it reasonable to think that these samples could have come from the above population?

d. What do you observe about the sample distributions as the sample size increases?

4. Based on your random sample of prices from Exercise 6, answer the following questions:

a. It looks like a lot of the prices end in 9. Do your sample results support that claim? Why or why not?

b. What is the typical price of the items in your sample? Explain how you found the price and why you chose that method.

EUREKA
MATH

5. The sample distributions of prices for three different random samples of 25 items from a grocery store are shown below.

    a. How do the distributions compare?

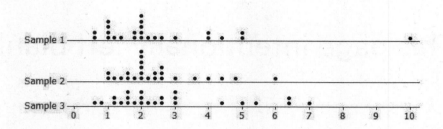

Dot Plots of Three Samples

    b. Thomas says that if he counts the items in his cart at that grocery store and multiplies by $2.00, he will have a pretty good estimate of how much he will have to pay. What do you think of his strategy?

This page intentionally left blank

Lesson 16: Methods for Selecting a Random Sample

Classwork

In this lesson, you will obtain random numbers to select a random sample. You will also design a plan for selecting a random sample to answer a statistical question about a population.

Example 1: Sampling Children's Books

What is the longest book you have ever read? *The Hobbit* has 95,022 words, and *The Cat in the Hat* has 830 words. Popular books vary in the number of words they have—not just the number of *different* words but the total number of words. The table below shows the total number of words in some of those books. The histogram displays the total number of words in 150 best-selling children's books with fewer than 100,000 words.

Book	Words	Book	Words	Book	Words
Black Beauty	59,635	Charlie and the Chocolate Factory	30,644	The Hobbit	95,022
The Catcher in the Rye	73,404	Old Yeller	35,968	Judy Moody Was in a Mood	11,049
The Adventures of Tom Sawyer	69,066	The Cat in the Hat	830	Treasure Island	66,950
The Secret Garden	80,398	Green Eggs and Ham	702	Magic Tree House Lions at Lunchtime	5,313
The Mouse and the Motorcycle	22,416	Little Bear	1,630	Harry Potter and the Sorcerer's Stone	77,325
The Wind in the Willows	58,424	The Red Badge of Courage	47,180	Harry Potter and the Chamber of Secrets	84,799
My Father's Dragon	7,682	Anne Frank: The Diary of a Young Girl	82,762	Junie B. Jones and the Stupid Smelly Bus	6,570
Frog and Toad All Year	1,727	Midnight for Charlie Bone	65,006	White Mountains	44,763
Book of Three	46,926	The Lion, The Witch and the Wardrobe	36,363	Double Fudge	38,860

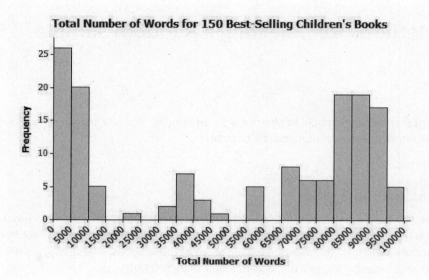

Total Number of Words for 150 Best-Selling Children's Books

Exercises 1–2

1. From the table, choose two books with which you are familiar, and describe their locations in the data distribution shown in the histogram.

2. Put dots on the number line below that you think would represent a random sample of size 10 from the number of words distribution above.

Lesson 16:   Methods for Selecting a Random Sample

Example 2: Using Random Numbers to Select a Sample

The histogram indicates the differences in the number of words in the collection of 150 books. How many words are typical for a best-selling children's book? Answering this question would involve collecting data, and there would be variability in those data. This makes the question a statistical question. Think about the 150 books used to create the histogram on the previous page as a population. How would you go about collecting data to determine the typical number of words for the books in this population?

How would you choose a random sample from the collection of 150 books discussed in this lesson?

The data for the number of words in the 150 best-selling children's books are listed below. Select a random sample of the number of words for 10 books.

Books 1–10	59,635	82,762	92,410	75,340	8,234	59,705	92,409	75,338	8,230	82,768
Books 11–20	73,404	65,006	88,250	2,100	81,450	72,404	88,252	2,099	81,451	65,011
Books 21–30	69,066	36,363	75,000	3,000	80,798	69,165	75,012	3,010	80,790	36,361
Books 31–40	80,398	95,022	71,200	3,250	81,450	80,402	71,198	3,252	81,455	95,032
Books 41–50	22,416	11,049	81,400	3,100	83,475	22,476	81,388	3,101	83,472	11,047
Books 51–60	58,424	66,950	92,400	2,750	9,000	58,481	92,405	2,748	9,002	66,954
Books 61–70	7,682	5,313	83,000	87,000	89,170	7,675	83,021	87,008	89,167	5,311
Books 71–80	1,727	77,325	89,010	862	88,365	1,702	89,015	860	88,368	77,328
Books 81–90	46,926	84,799	88,045	927	89,790	46,986	88,042	926	89,766	84,796
Books 91–100	30,644	6,570	90,000	8,410	91,010	30,692	90,009	8,408	91,015	6,574
Books 101–110	35,968	44,763	89,210	510	9,247	35,940	89,213	512	9,249	44,766
Books 111–120	830	8,700	92,040	7,891	83,150	838	92,037	7,889	83,149	8,705
Books 121–130	702	92,410	94,505	38,860	81,110	712	94,503	87,797	81,111	92,412
Books 131–140	1,630	88,250	97,000	7,549	8,245	1,632	97,002	7,547	8,243	88,254
Books 141–150	47,180	75,000	89,241	81,234	8,735	47,192	89,239	81,238	8,739	75,010

Exercises 3–6

3. Follow your teacher's instructions to generate a set of 10 random numbers. Find the total number of words corresponding to each book identified by your random numbers.

4. Choose two more different random samples of size 10 from the data, and make a dot plot of each of the three samples.

5. If your teacher randomly chooses 10 books from your summer vacation reading list, would you be likely to get many books with a lot of words? Explain your thinking using statistical terms.

6. If you were to compare your samples to your classmates' samples, do you think your answer to Exercise 5 would change? Why or why not?

Exercises 7–9: A Statistical Study of Balance and Grade

7. Is the following question a statistical question: Do sixth graders or seventh graders tend to have better balance?

Lesson 16:   Methods for Selecting a Random Sample

8. Berthio's class decided to measure balance by finding out how long people can stand on one foot.

   a. How would you rephrase the question from Exercise 7 to create a statistical question using this definition of balance? Explain your reasoning.

   b. What should the class think about to be consistent in how they collect the data if they actually have people stand on one foot and measure the time?

9. Work with your class to devise a plan to select a random sample of sixth graders and a random sample of seventh graders to measure their balance using Berthio's method. Then, write a paragraph describing how you will collect data to determine whether there is a difference in how long sixth graders and seventh graders can stand on one foot. Your plan should answer the following questions:

   a. What is the population? How will samples be selected from the population? Why is it important that they be random samples?

   b. How would you conduct the activity?

   c. What sample statistics will you calculate, and how will you display and analyze the data?

d.   What would you accept as evidence that there actually is a difference in how long sixth graders can stand on one foot compared to seventh graders?

Problem Set

1. The suggestions below for how to choose a random sample of students at your school were made and vetoed. Explain why you think each was vetoed.

   a. Use every fifth person you see in the hallway before class starts.

   b. Use all of the students taking math the same time that your class meets.

   c. Have students who come to school early do the activity before school starts.

   d. Have everyone in the class find two friends to be in the sample.

2. A teacher decided to collect homework from a random sample of her students rather than grading every paper every day.

   a. Describe how she might choose a random sample of five students from her class of 35 students.

   b. Suppose every day for 75 days throughout an entire semester she chooses a random sample of five students. Do you think some students will never get selected? Why or why not?

3. Think back to earlier lessons in which you chose a random sample. Describe how you could have used a random number generator to select a random sample in each case.

   a. A random sample of the words in the poem "Casey at the Bat"

   b. A random sample of the grocery prices on a weekly flyer

4. Sofia decided to use a different plan for selecting a random sample of books from the population of 150 top-selling children's books from Example 2. She generated ten random numbers between 1 and 100,000 to stand for the possible number of pages in any of the books. Then, she found the books that had the number of pages specified in the sample. What would you say to Sofia?

5. Find an example from a newspaper, a magazine, or another source that used a sample. Describe the population, the sample, the sample statistic, how you think the sample might have been chosen, and whether or not you think the sample was random.

This page intentionally left blank

Lesson 17: Sampling Variability

Classwork

Example 1: Estimating a Population Mean

The owners of a gym have been keeping track of how long each person spends at the gym. Eight hundred of these times (in minutes) are shown in the population tables located at the end of the lesson. These 800 times will form the *population* that you will investigate in this lesson.

Look at the values in the population. Can you find the longest time spent in the gym in the population? Can you find the shortest?

On average, roughly how long do you think people spend at the gym? In other words, by just looking at the numbers in the two tables, make an estimate of the *population mean*.

You could find the population mean by typing all 800 numbers into a calculator or a computer, adding them up, and dividing by 800. This would be extremely time consuming, and usually it is not possible to measure every value in a population.

Instead of doing a calculation using every value in the population, we will use a *random sample* to find the mean of the sample. The sample mean will then be used as an estimate of the population mean.

Example 2: Selecting a Sample Using a Table of Random Digits

The table of random digits provided with this lesson will be used to select items from a population to produce a random sample from the population. The list of digits is determined by a computer program that simulates a random selection of the digits 0, 1, 2, 3, 4, 5, 6, 7, 8, or 9. Imagine that each of these digits is written on a slip of paper and placed in a bag. After thoroughly mixing the bag, one slip is drawn, and its digit is recorded in this list of random digits. The slip is then returned to the bag, and another slip is selected. The digit on this slip is recorded and then returned to the bag. The process is repeated over and over. The resulting list of digits is called a *random number table*.

How could you use a table of random digits to take a random sample?

Step 1: Place the table of random digits in front of you. Without looking at the page, place the eraser end of your pencil somewhere on the table. Start using the table of random digits at the number closest to where your eraser touched the paper. This digit and the following two specify which observation from the population tables will be the first observation in your sample.

For example, suppose the eraser end of your pencil lands on the twelfth number in Row 3 of the random digit table. This number is 5, and the two following numbers are 1 and 4. This means that the first observation in your sample is observation number 514 from the population. Find observation number 514 in the population table. Do this by going to Row 51 and moving across to the column heading "4." This observation is 53, so the first observation in your sample is 53.

If the number from the random number table is any number 800 or greater, you will ignore this number and use the next three digits in the table.

Step 2: Continue using the table of random digits from the point you reached, and select the other four observations in your sample like you did in Step 1.

For example, continuing on from the position in the example given in Step 1:

- The next number from the random digit table is 716, and observation 716 is 63.
- The next number from the random digit table is 565, and observation 565 is 31.
- The next number from the random digit table is 911, and there is no observation 911. So, we ignore these three digits.
- The next number from the random digit table is 928, and there is no observation 928. So, we ignore these three digits.
- The next number from the random digit table is 303, and observation 303 is 70.
- The next number from the random digit table is 677, and observation 677 is 42.

Exercises 1–4

Initially, you will select just five values from the population to form your sample. This is a very small sample size, but it is a good place to start to understand the ideas of this lesson.

1. Use the table of random numbers to select five values from the population of times. What are the five observations in your sample?

2. For the sample that you selected, calculate the sample mean.

3. You selected a random sample and calculated the sample mean in order to estimate the population mean. Do you think that the mean of these five observations is exactly correct for the population mean? Could the population mean be greater than the number you calculated? Could the population mean be less than the number you calculated?

4. In practice, you only take one sample in order to estimate a population characteristic. But, for the purposes of this lesson, suppose you were to take another random sample from the same population of times at the gym. Could the new sample mean be closer to the population mean than the mean of these five observations? Could it be farther from the population mean?

Exercises 5–7

As a class, you will now investigate sampling variability by taking several samples from the same population. Each sample will have a different sample mean. This variation provides an example of sampling variability.

5. Place the table of random digits in front of you, and without looking at the page, place the eraser end of your pencil somewhere on the table of random numbers. Start using the table of random digits at the number closest to where your eraser touches the paper. This digit and the following two specify which observation from the population tables will be the first observation in your sample. Write this three-digit number and the corresponding data value from the population in the space below.

6. Continue moving to the right in the table of random digits from the place you ended in Exercise 5. Use three digits at a time. Each set of three digits specifies which observation in the population is the next number in your sample. Continue until you have four more observations, and write these four values in the space below.

7. Calculate the mean of the five values that form your sample. Round your answer to the nearest tenth. Show your work and your sample mean in the space below.

Exercises 8–11

You will now use the sample means from Exercise 7 from the entire class to make a dot plot.

8. Write the sample means for everyone in the class in the space below.

9. Use all the sample means to make a dot plot using the axis given below. (Remember, if you have repeated or close values, stack the dots one above the other.)

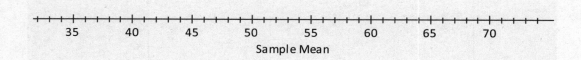

Sample Mean

10. What do you see in the dot plot that demonstrates sampling variability?

11. Remember that in practice you only take one sample. (In this lesson, many samples were taken in order to demonstrate the concept of sampling variability.) Suppose that a statistician plans to take a random sample of size 5 from the population of times spent at the gym and that he will use the sample mean as an estimate of the population mean. Approximately how far can the statistician expect the sample mean to be from the population mean?

Population

	0	1	2	3	4	5	6	7	8	9
00	45	58	49	78	59	36	52	39	70	51
01	50	45	45	66	71	55	65	33	60	51
02	53	83	40	51	83	57	75	38	43	77
03	49	49	81	57	42	36	22	66	68	52
04	60	67	43	60	55	63	56	44	50	58
05	64	41	67	73	55	69	63	46	50	65
06	54	58	53	55	51	74	53	55	64	16
07	28	48	62	24	82	51	64	45	41	47
08	70	50	38	16	39	83	62	50	37	58
09	79	62	45	48	42	51	67	68	56	78
10	61	56	71	55	57	77	48	65	61	62
11	65	40	56	47	44	51	38	68	64	40
12	53	22	73	62	82	78	84	50	43	43
13	81	42	72	49	55	65	41	92	50	60
14	56	44	40	70	52	47	30	9	58	53
15	84	64	64	34	37	69	57	75	62	67
16	45	58	49	78	59	36	52	39	70	51
17	50	45	45	66	71	55	65	33	60	51
18	53	83	40	51	83	57	75	38	43	77
19	49	49	81	57	42	36	22	66	68	52
20	60	67	43	60	55	63	56	44	50	58
21	64	41	67	73	55	69	63	46	50	65
22	54	58	53	55	51	74	53	55	64	16
23	28	48	62	24	82	51	64	45	41	47
24	70	50	38	16	39	83	62	50	37	58
25	79	62	45	48	42	51	67	68	56	78
26	61	56	71	55	57	77	48	65	61	62
27	65	40	56	47	44	51	38	68	64	40
28	53	22	73	62	82	78	84	50	43	43
29	81	42	72	49	55	65	41	92	50	60
30	56	44	40	70	52	47	30	9	58	53
31	84	64	64	34	37	69	57	75	62	67
32	45	58	49	78	59	36	52	39	70	51
33	50	45	45	66	71	55	65	33	60	51
34	53	83	40	51	83	57	75	38	43	77
35	49	49	81	57	42	36	22	66	68	52
36	60	67	43	60	55	63	56	44	50	58
37	64	41	67	73	55	69	63	46	50	65
38	54	58	53	55	51	74	53	55	64	16
39	28	48	62	24	82	51	64	45	41	47

Population (continued)

	0	1	2	3	4	5	6	7	8	9
40	53	70	59	62	33	31	74	44	46	68
41	37	51	84	47	46	33	53	54	70	74
42	35	45	48	45	56	60	66	60	65	57
43	42	81	67	64	60	79	46	48	67	56
44	41	21	41	58	48	38	50	53	73	38
45	35	28	43	43	55	39	75	45	68	36
46	64	31	31	40	84	79	47	63	48	46
47	34	36	54	61	33	16	50	60	52	55
48	53	52	48	47	77	37	66	51	61	64
49	40	44	45	22	36	64	50	49	64	39
50	45	69	67	33	55	61	62	38	51	43
51	55	39	46	56	53	50	44	42	40	60
52	11	36	56	69	72	73	71	48	58	52
53	81	47	36	54	81	59	50	42	80	69
54	40	43	30	54	61	13	73	65	52	40
55	71	78	71	61	54	79	63	47	49	73
56	53	70	59	62	33	31	74	44	46	68
57	37	51	84	47	46	33	53	54	70	74
58	35	45	48	45	56	60	66	60	65	57
59	42	81	67	64	60	79	46	48	67	56
60	41	21	41	58	48	38	50	53	73	38
61	35	28	43	43	55	39	75	45	68	36
62	64	31	31	40	84	79	47	63	48	46
63	34	36	54	61	33	16	50	60	52	55
64	53	52	48	47	77	37	66	51	61	64
65	40	44	45	22	36	64	50	49	64	39
66	45	69	67	33	55	61	62	38	51	43
67	55	39	46	56	53	50	44	42	40	60
68	11	36	56	69	72	73	71	48	58	52
69	81	47	36	54	81	59	50	42	80	69
70	40	43	30	54	61	13	73	65	52	40
71	71	78	71	61	54	79	63	47	49	73
72	53	70	59	62	33	31	74	44	46	68
73	37	51	84	47	46	33	53	54	70	74
74	35	45	48	45	56	60	66	60	65	57
75	42	81	67	64	60	79	46	48	67	56
76	41	21	41	58	48	38	50	53	73	38
77	35	28	43	43	55	39	75	45	68	36
78	64	31	31	40	84	79	47	63	48	46
79	34	36	54	61	33	16	50	60	52	55

Lesson 17:   Sampling Variability

Table of Random Digits

Row																				
1	6	6	7	2	8	0	0	8	4	0	0	4	6	0	3	2	2	4	6	8
2	8	0	3	1	1	1	1	2	7	0	1	9	1	2	7	1	3	3	5	3
3	5	3	5	7	3	6	3	1	7	2	5	5	1	4	7	1	6	5	6	5
4	9	1	1	9	2	8	3	0	3	6	7	7	4	7	5	9	8	1	8	3
5	9	0	2	9	9	7	4	6	3	6	6	3	7	4	2	7	0	0	1	9
6	8	1	4	6	4	6	8	2	8	9	5	5	2	9	6	2	5	3	0	3
7	4	1	1	9	7	0	7	2	9	0	9	7	0	4	6	2	3	1	0	9
8	9	9	2	7	1	3	2	9	0	3	9	0	7	5	6	7	1	7	8	7
9	3	4	2	2	9	1	9	0	7	8	1	6	2	5	3	9	0	9	1	0
10	2	7	3	9	5	9	9	3	2	9	3	9	1	9	0	5	5	1	4	2
11	0	2	5	4	0	8	1	7	0	7	1	3	0	4	3	0	6	4	4	4
12	8	6	0	5	4	8	8	2	7	7	0	1	0	1	7	1	3	5	3	4
13	4	2	6	4	5	2	4	2	6	1	7	5	6	6	4	0	8	4	1	2
14	4	4	9	8	7	3	4	3	8	2	9	1	5	3	5	9	8	9	2	9
15	6	4	8	0	0	0	4	2	3	8	1	8	4	0	9	5	0	9	0	4
16	3	2	3	8	4	8	8	6	2	9	1	0	1	9	9	3	0	7	3	5
17	6	6	7	2	8	0	0	8	4	0	0	4	6	0	3	2	2	4	6	8
18	8	0	3	1	1	1	1	2	7	0	1	9	1	2	7	1	3	3	5	3
19	5	3	5	7	3	6	3	1	7	2	5	5	1	4	7	1	6	5	6	5
20	9	1	1	9	2	8	3	0	3	6	7	7	4	7	5	9	8	1	8	3
21	9	0	2	9	9	7	4	6	3	6	6	3	7	4	2	7	0	0	1	9
22	8	1	4	6	4	6	8	2	8	9	5	5	2	9	6	2	5	3	0	3
23	4	1	1	9	7	0	7	2	9	0	9	7	0	4	6	2	3	1	0	9
24	9	9	2	7	1	3	2	9	0	3	9	0	7	5	6	7	1	7	8	7
25	3	4	2	2	9	1	9	0	7	8	1	6	2	5	3	9	0	9	1	0
26	2	7	3	9	5	9	9	3	2	9	3	9	1	9	0	5	5	1	4	2
27	0	2	5	4	0	8	1	7	0	7	1	3	0	4	3	0	6	4	4	4
28	8	6	0	5	4	8	8	2	7	7	0	1	0	1	7	1	3	5	3	4
29	4	2	6	4	5	2	4	2	6	1	7	5	6	6	4	0	8	4	1	2
30	4	4	9	8	7	3	4	3	8	2	9	1	5	3	5	9	8	9	2	9
31	6	4	8	0	0	0	4	2	3	8	1	8	4	0	9	5	0	9	0	4
32	3	2	3	8	4	8	8	6	2	9	1	0	1	9	9	3	0	7	3	5
33	6	6	7	2	8	0	0	8	4	0	0	4	6	0	3	2	2	4	6	8
34	8	0	3	1	1	1	1	2	7	0	1	9	1	2	7	1	3	3	5	3
35	5	3	5	7	3	6	3	1	7	2	5	5	1	4	7	1	6	5	6	5
36	9	1	1	9	2	8	3	0	3	6	7	7	4	7	5	9	8	1	8	3
37	9	0	2	9	9	7	4	6	3	6	6	3	7	4	2	7	0	0	1	9
38	8	1	4	6	4	6	8	2	8	9	5	5	2	9	6	2	5	3	0	3
39	4	1	1	9	7	0	7	2	9	0	9	7	0	4	6	2	3	1	0	9
40	9	9	2	7	1	3	2	9	0	3	9	0	7	5	6	7	1	7	8	7

Lesson Summary

A population characteristic is estimated by taking a random sample from the population and calculating the value of a statistic for the sample. For example, a population mean is estimated by selecting a random sample from the population and calculating the sample mean.

The value of the sample statistic (e.g., the sample mean) will vary based on the random sample that is selected. This variation from sample to sample in the values of the sample statistic is called *sampling variability*.

Problem Set

1. Yousef intends to buy a car. He wishes to estimate the mean fuel efficiency (in miles per gallon) of all cars available at this time. Yousef selects a random sample of 10 cars and looks up their fuel efficiencies on the Internet. The results are shown below.

   22 25 29 23 31 29 28 22 23 27

   a. Yousef will estimate the mean fuel efficiency of all cars by calculating the mean for his sample. Calculate the sample mean, and record your answer. (Be sure to show your work.)

   b. In practice, you only take one sample to estimate a population characteristic. However, if Yousef were to take another random sample of 10 cars from the same population, would he likely get the same value for the sample mean?

   c. What if Yousef were to take *many* random samples of 10 cars? Would all of the sample means be the same?

   d. Using this example, explain what sampling variability means.

2. Think about the mean number of siblings (brothers and sisters) for all students at your school.

   a. What do you think is the approximate value of the mean number of siblings for the population of all students at your school?

   b. How could you find a better estimate of this population mean?

   c. Suppose that you have now selected a random sample of students from your school. You have asked all of the students in your sample how many siblings they have. How will you calculate the sample mean?

   d. If you had taken a different sample, would the sample mean have taken the same value?

   e. There are many different samples of students that you could have selected. These samples produce many different possible sample means. What is the phrase used for this concept?

   f. Does the phrase you gave in part (e) apply only to sample means?

Lesson 18: Sampling Variability and the Effect of Sample Size

Classwork

Example 1: Sampling Variability

The previous lesson investigated the statistical question "What is the typical time spent at the gym?" by selecting random samples from the population of 800 gym members. Two different dot plots of sample means calculated from random samples from the population are displayed below. The first dot plot represents the means of 20 samples with each sample having 5 data points. The second dot plot represents the means of 20 samples with each sample having 15 data points.

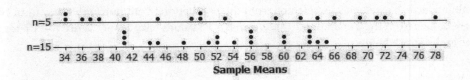

Based on the first dot plot, Jill answered the statistical question by indicating the mean time people spent at the gym was between 34 and 78 minutes. She decided that a time approximately in the middle of that interval would be her estimate of the mean time the 800 people spent at the gym. She estimated 52 minutes. Scott answered the question using the second dot plot. He indicated that the mean time people spent at the gym was between 41 and 65 minutes. He also selected a time of 52 minutes to answer the question.

a.   Describe the differences in the two dot plots.

b.   Which dot plot do you feel more confident in using to answer the statistical question? Explain your answer.

c.   In general, do you want sampling variability to be large or small? Explain.

Exercises 1–3

In the previous lesson, you saw a population of 800 times spent at the gym. You will now select a random sample of size 15 from that population. You will then calculate the sample mean.

1. Start by selecting a three-digit number from the table of random digits. Place the random digit table in front of you. Without looking at the page, place the eraser end of your pencil somewhere on the table of random digits. Start using the table of random digits at the digit closest to your eraser. This digit and the following two specify which observation from the population will be the first observation in your sample. Write the value of this observation in the space below. (Discard any three-digit number that is 800 or larger, and use the next three digits from the random digit table.)

2. Continue moving to the right in the table of random digits from the point that you reached in Exercise 1. Each three-digit number specifies a value to be selected from the population. Continue in this way until you have selected 14 more values from the population. This will make 15 values altogether. Write the values of all 15 observations in the space below.

3. Calculate the mean of your 15 sample values. Write the value of your sample mean below. Round your answer to the nearest tenth. (Be sure to show your work.)

Exercises 4–6

You will now use the sample means from Exercise 3 for the entire class to make a dot plot.

4. Write the sample means for everyone in the class in the space below.

EUREKA
MATH

5. Use all the sample means to make a dot plot using the axis given below. (Remember, if you have repeated values or values close to each other, stack the dots one above the other.)

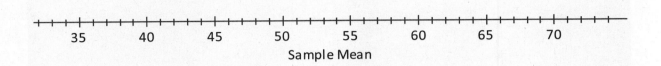

6. In the previous lesson, you drew a dot plot of sample means for samples of size 5. How does the dot plot above (of sample means for samples of size 15) compare to the dot plot of sample means for samples of size 5? For which sample size (5 or 15) does the sample mean have the greater sampling variability?

    This exercise illustrates the notion that the greater the sample size, the smaller the sampling variability of the sample mean.

Exercises 7–8

7. Remember that in practice you only take one sample. Suppose that a statistician plans to take a random sample of size 15 from the population of times spent at the gym and will use the sample mean as an estimate of the population mean. Based on the dot plot of sample means that your class collected from the population, approximately how far can the statistician expect the sample mean to be from the population mean? (The actual population mean is 53.9 minutes.)

8. How would your answer in Exercise 7 compare to the equivalent mean of the distances for a sample of size 5?

Exercises 9–11

Suppose everyone in your class selected a random sample of size 25 from the population of times spent at the gym.

9. What do you think the dot plot of the class's sample means would look like? Make a sketch using the axis below.

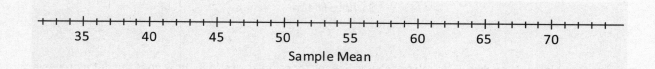

10. Suppose that a statistician plans to estimate the population mean using a sample of size 25. According to your sketch, approximately how far can the statistician expect the sample mean to be from the population mean?

11. Suppose you have a choice of using a sample of size 5, 15, or 25. Which of the three makes the sampling variability of the sample mean the smallest? Why would you choose the sample size that makes the sampling variability of the sample mean as small as possible?

Problem Set

1. The owner of a new coffee shop is keeping track of how much each customer spends (in dollars). One hundred of these amounts are shown in the table below. These amounts will form the *population* for this question.

	0	1	2	3	4	5	6	7	8	9
0	6.18	4.67	4.01	4.06	3.28	4.47	4.86	4.91	3.96	6.18
1	4.98	5.42	5.65	2.97	2.92	7.09	2.78	4.20	5.02	4.98
2	3.12	1.89	4.19	5.12	4.38	5.34	4.22	4.27	5.25	3.12
3	3.90	4.47	4.07	4.80	6.28	5.79	6.07	7.64	6.33	3.90
4	5.55	4.99	3.77	3.63	5.21	3.85	7.43	4.72	6.53	5.55
5	4.55	5.38	5.83	4.10	4.42	5.63	5.57	5.32	5.32	4.55
6	4.56	7.67	6.39	4.05	4.51	5.16	5.29	6.34	3.68	4.56
7	5.86	4.75	4.94	3.92	4.84	4.95	4.50	4.56	7.05	5.86
8	5.00	5.47	5.00	5.70	5.71	6.19	4.41	4.29	4.34	5.00
9	5.12	5.58	6.16	6.39	5.93	3.72	5.92	4.82	6.19	5.12

a. Place the table of random digits in front of you. Select a starting point without looking at the page. Then, taking two digits at a time, select a random sample of size 10 from the population above. Write the 10 values in the space below. (For example, suppose you start at the third digit of row four of the random digit table. Taking two digits gives you 19. In the population above, go to the row labeled 1, and move across to the column labeled 9. This observation is 4.98, and that will be the first observation in your sample. Then, continue in the random digit table from the point you reached.)

Calculate the mean for your sample, showing your work. Round your answer to the nearest thousandth.

b. Using the same approach as in part (a), select a random sample of size 20 from the population.

Calculate the mean for your sample of size 20. Round your answer to the nearest thousandth.

c. Which of your sample means is likely to be the better estimate of the population mean? Explain your answer in terms of sampling variability.

2. Two dot plots are shown below. One of the dot plots shows the values of some sample means from random samples of size 10 from the population given in Problem 1. The other dot plot shows the values of some sample means from random samples of size 20 from the population given in Problem 1.

Dot Plot A

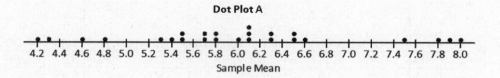

Sample Mean

Dot Plot B

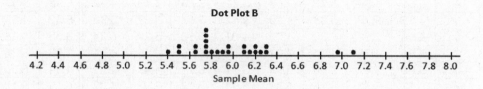

Sample Mean

Which dot plot is for sample means from samples of size 10, and which dot plot is for sample means from samples of size 20? Explain your reasoning.

The sample means from samples of size 10 are shown in Dot Plot _____.

The sample means from samples of size 20 are shown in Dot Plot _____.

3. You are going to use a random sample to estimate the mean travel time for getting to school for all the students in your grade. You will select a random sample of students from your grade. Explain why you would like the sampling variability of the sample mean to be *small*.

EUREKA
MATH™

Lesson 19: Understanding Variability When Estimating a Population Proportion

Classwork

In a previous lesson, you selected several random samples from a population. You recorded values of a numerical variable. You then calculated the mean for each sample, saw that there was variability in the sample means, and created a distribution of sample means to better see the sampling variability. You then considered larger samples and saw that the variability in the distribution decreases when the sample size increases. In this lesson, you will use a similar process to investigate variability in sample proportions.

Example 1: Sample Proportion

Your teacher will give your group a bag that contains colored cubes, some of which are red. With your classmates, you are going to build a distribution of sample proportions.

a. Each person in your group should randomly select a sample of 10 cubes from the bag. Record the data for your sample in the table below.

Cube	Outcome (Color)
1	
2	
3	
4	
5	
6	
7	
8	
9	
10	

b. What is the proportion of red cubes in your sample of 10?

This value is called the *sample proportion*. The sample proportion is found by dividing the number of successes (in this example, the number of red cubes) by the total number of observations in the sample.

c. Write your sample proportion on a sticky note, and place it on the number line that your teacher has drawn on the board. Place your note above the value on the number line that corresponds to your sample proportion.

The graph of all students' sample proportions is called a *sampling distribution* of the sample proportions.

d. Describe the shape of the distribution.

e. Describe the variability in the sample proportions.

Based on the distribution, answer the following:

f. What do you think is the population proportion?

g. How confident are you of your estimate?

Example 2: Sampling Variability

What do you think would happen to the sampling distribution if everyone in class took a random sample of 30 cubes from the bag? To help answer this question, you will repeat the random sampling you did in part (a) of Example 1, except now you will draw a random sample of 30 cubes instead of 10.

a. Take a random sample of 30 cubes from the bag. Carefully record the outcome of each draw.

Cube	Outcome (Color)	Cube	Outcome (Color)
1		16	
2		17	
3		18	
4		19	
5		20	
6		21	
7		22	
8		23	
9		24	
10		25	
11		26	
12		27	
13		28	
14		29	
15		30	

b. What is the proportion of red cubes in your sample of 30?

c. Write your sample proportion on a sticky note, and place the note on the number line that your teacher has drawn on the board. Place your note above the value on the number line that corresponds to your sample proportion.

d. Describe the shape of the distribution.

Exercises 1–5

1. Describe the variability in the sample proportions.

2. Based on the distribution, answer the following:

    a. What do you think is the population proportion?

    b. How confident are you of your estimate?

    c. If you were taking a random sample of 30 cubes and determined the proportion that was red, do you think your sample proportion will be within 0.05 of the population proportion? Explain.

3. Compare the sampling distribution based on samples of size 10 to the sampling distribution based on samples of size 30.

4. As the sample size increased from 10 to 30, describe what happened to the sampling variability of the sample proportions.

5. What do you think would happen to the sampling variability of the sample proportions if the sample size for each sample was 50 instead of 30? Explain.

Problem Set

1. A class of seventh graders wanted to find the proportion of M&M's® that are red. Each seventh grader took a random sample of 20 M&M's® from a very large container of M&M's®. The following is the proportion of red M&M's each student found.

0.15	0	0.1	0.1	0.05	0.1	0.2	0.05	0.1
0.1	0.15	0.2	0	0.1	0.15	0.15	0.1	0.2
0.3	0.1	0.1	0.2	0.1	0.15	0.1	0.05	0.3

   a. Construct a dot plot of the sample proportions.

   b. Describe the shape of the distribution.

   c. Describe the variability of the distribution.

   d. Suppose the seventh-grade students had taken random samples of size 50. Describe how the sampling distribution would change from the one you constructed in part (a).

2. A group of seventh graders wanted to estimate the proportion of middle school students who suffer from allergies. The members of one group of seventh graders each took a random sample of 10 middle school students, and the members of another group of seventh graders each took a random sample of 40 middle school students. Below are two sampling distributions of the sample proportions of middle school students who said that they suffer from allergies. Which dot plot is based on random samples of size 40? How can you tell?

Dot Plot A:

Dot Plot of Sample Proportion

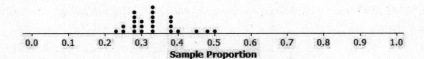

Dot Plot B:

Dot Plot of Sample Proportion

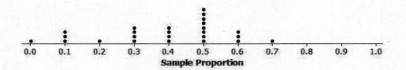

3. The nurse in your school district would like to study the proportion of middle school students who usually get at least eight hours of sleep on school nights. Suppose each student in your class plans on taking a random sample of 20 middle school students from your district, and each calculates a sample proportion of students who said that they usually get at least eight hours of sleep on school nights.

    a. Do you expect everyone in your class to get the same value for their sample proportions? Explain.

    b. Suppose each student in class increased the sample size from 20 to 40. Describe how you could reduce the sampling variability.

EUREKA
MATH™

Lesson 20: Estimating a Population Proportion

Classwork

In a previous lesson, each student in your class selected a random sample from a population and calculated the sample proportion. It was observed that there was sampling variability in the sample proportions, and as the sample size increased, the variability decreased. In this lesson, you will investigate how sample proportions can be used to estimate population proportions.

Example 1: Mean of Sample Proportions

A class of 30 seventh graders wanted to estimate the proportion of middle school students who were vegetarians. Each seventh grader took a random sample of 20 middle school students. Students were asked the question, "Are you a vegetarian?" One sample of 20 students had three students who said that they were vegetarians. For this sample, the sample proportion is $\frac{3}{20}$, or 0.15. The following are the proportions of vegetarians the seventh graders found in 30 samples. Each sample was of size 20 students. The proportions are rounded to the nearest hundredth.

0.15	0.10	0.15	0.00	0.10	0.15	0.10	0.10	0.05	0.20
0.25	0.15	0.25	0.25	0.30	0.20	0.10	0.20	0.05	0.10
0.10	0.30	0.15	0.05	0.25	0.15	0.20	0.10	0.20	0.15

Exercises 1–9

1. The first student reported a sample proportion of 0.15. Interpret this value in terms of the summary of the problem in the example.

2. Another student reported a sample proportion of 0. Did this student do something wrong when selecting the sample of middle school students?

3. Assume you were part of this seventh-grade class and you got a sample proportion of 0.20 from a random sample of middle school students. Based on this sample proportion, what is your estimate for the proportion of all middle school students who are vegetarians?

4. Construct a dot plot of the 30 sample proportions.

5. Describe the shape of the distribution.

6. Using the 30 class results listed on the previous page, what is your estimate for the proportion of all middle school students who are vegetarians? Explain how you made this estimate.

7. Calculate the mean of the 30 sample proportions. How close is this value to the estimate you made in Exercise 6?

EUREKA MATH™

8. The proportion of all middle school students who are vegetarians is 0.15. This is the actual proportion for the entire population of middle school students used to select the samples. How the mean of the 30 sample proportions compares with the actual population proportion depends on the students' samples.

9. Do the sample proportions in the dot plot tend to cluster around the value of the population proportion? Are any of the sample proportions far away from 0.15? List the proportions that are far away from 0.15.

Example 2: Estimating Population Proportion

Two hundred middle school students at Roosevelt Middle School responded to several survey questions. A printed copy of the responses the students gave to various questions will be provided by your teacher.

The data are organized in columns and are summarized by the following table:

Column Heading	Description
ID	Numbers from 1 to 200
Travel to School	Method used to get to school: Walk, car, rail, bus, bicycle, skateboard/scooter/rollerblade, boat
Favorite Season	Summer, fall, winter, spring
Allergies	Yes or no
Favorite School Subject	Art, English, languages, social studies, history, geography, music, science, computers, math, PE, other
Favorite Music	Classical, country, heavy metal, jazz, pop, punk rock, rap, reggae, R&B, rock and roll, techno, gospel, other
What superpower would you like?	Invisibility, super strength, telepathy, fly, freeze time

The last column in the data file is based on the question: Which of the following superpowers would you most like to have? The choices were invisibility, super strength, telepathy, fly, or freeze time.

The class wants to determine the proportion of Roosevelt Middle School students who answered "freeze time" to the last question. You will use a sample of the Roosevelt Middle School population to estimate the proportion of the students who answered "freeze time" to the last question.

A random sample of 20 student responses is needed. You are provided the random number table you used in a previous lesson. A printed list of the 200 Roosevelt Middle School students is also provided. In small groups, complete the following exercise:

    a.   Select a random sample of 20 student responses from the data file. Explain how you selected the random sample.

    b.   In the table below, list the 20 responses for your sample.

	Response
1	
2	
3	
4	
5	
6	
7	
8	
9	
10	
11	
12	
13	
14	
15	
16	
17	
18	
19	
20	

    c.   Estimate the population proportion of students who responded "freeze time" by calculating the sample proportion of the 20 sampled students who responded "freeze time" to the question.

      Lesson 20:   Estimating a Population Proportion

EUREKA
MATH™

d.   Combine your sample proportion with other students' sample proportions, and create a dot plot of the distribution of the sample proportions of students who responded "freeze time" to the question.

e.   By looking at the dot plot, what is the value of the proportion of the 200 Roosevelt Middle School students who responded "freeze time" to the question?

f.   Usually, you will estimate the proportion of Roosevelt Middle School students using just a single sample proportion. How different was your sample proportion from your estimate based on the dot plot of many samples?

g.   Circle your sample proportion on the dot plot. How does your sample proportion compare with the mean of all the sample proportions?

h.   Calculate the mean of all of the sample proportions. Locate the mean of the sample proportions in your dot plot; mark this position with an X. How does the mean of the sample proportions compare with your sample proportion?

Problem Set

1. A class of 30 seventh graders wanted to estimate the proportion of middle school students who played a musical instrument. Each seventh grader took a random sample of 25 middle school students and asked each student whether or not she played a musical instrument. The following are the sample proportions the seventh graders found in 30 samples.

0.80	0.64	0.72	0.60	0.60	0.72	0.76	0.68	0.72	0.68
0.72	0.68	0.68	0.76	0.84	0.60	0.80	0.72	0.76	0.80
0.76	0.60	0.80	0.84	0.68	0.68	0.70	0.68	0.64	0.72

a. The first student reported a sample proportion of 0.80. What does this value mean in terms of this scenario?

b. Construct a dot plot of the 30 sample proportions.

c. Describe the shape of the distribution.

d. Describe the variability of the distribution.

e. Using the 30 class sample proportions listed above, what is your estimate for the proportion of all middle school students who played a musical instrument?

2. Select another variable or column from the data file that is of interest. Take a random sample of 30 students from the list, and record the response to your variable of interest of each of the 30 students.

a. Based on your random sample, what is your estimate for the proportion of all middle school students?

b. If you selected a second random sample of 30, would you get the same sample proportion for the second random sample that you got for the first random sample? Explain why or why not.

Table of Random Digits

Row																				
1	6	6	7	2	8	0	0	8	4	0	0	4	6	0	3	2	2	4	6	8
2	8	0	3	1	1	1	1	2	7	0	1	9	1	2	7	1	3	3	5	3
3	5	3	5	7	3	6	3	1	7	2	5	5	1	4	7	1	6	5	6	5
4	9	1	1	9	2	8	3	0	3	6	7	7	4	7	5	9	8	1	8	3
5	9	0	2	9	9	7	4	6	3	6	6	3	7	4	2	7	0	0	1	9
6	8	1	4	6	4	6	8	2	8	9	5	5	2	9	6	2	5	3	0	3
7	4	1	1	9	7	0	7	2	9	0	9	7	0	4	6	2	3	1	0	9
8	9	9	2	7	1	3	2	9	0	3	9	0	7	5	6	7	1	7	8	7
9	3	4	2	2	9	1	9	0	7	8	1	6	2	5	3	9	0	9	1	0
10	2	7	3	9	5	9	9	3	2	9	3	9	1	9	0	5	5	1	4	2
11	0	2	5	4	0	8	1	7	0	7	1	3	0	4	3	0	6	4	4	4
12	8	6	0	5	4	8	8	2	7	7	0	1	0	1	7	1	3	5	3	4
13	4	2	6	4	5	2	4	2	6	1	7	5	6	6	4	0	8	4	1	2
14	4	4	9	8	7	3	4	3	8	2	9	1	5	3	5	9	8	9	2	9
15	6	4	8	0	0	0	4	2	3	8	1	8	4	0	9	5	0	9	0	4
16	3	2	3	8	4	8	8	6	2	9	1	0	1	9	9	3	0	7	3	5
17	6	6	7	2	8	0	0	8	4	0	0	4	6	0	3	2	2	4	6	8
18	8	0	3	1	1	1	1	2	7	0	1	9	1	2	7	1	3	3	5	3
19	5	3	5	7	3	6	3	1	7	2	5	5	1	4	7	1	6	5	6	5
20	9	1	1	9	2	8	3	0	3	6	7	7	4	7	5	9	8	1	8	3
21	9	0	2	9	9	7	4	6	3	6	6	3	7	4	2	7	0	0	1	9
22	8	1	4	6	4	6	8	2	8	9	5	5	2	9	6	2	5	3	0	3
23	4	1	1	9	7	0	7	2	9	0	9	7	0	4	6	2	3	1	0	9
24	9	9	2	7	1	3	2	9	0	3	9	0	7	5	6	7	1	7	8	7
25	3	4	2	2	9	1	9	0	7	8	1	6	2	5	3	9	0	9	1	0
26	2	7	3	9	5	9	9	3	2	9	3	9	1	9	0	5	5	1	4	2
27	0	2	5	4	0	8	1	7	0	7	1	3	0	4	3	0	6	4	4	4
28	8	6	0	5	4	8	8	2	7	7	0	1	0	1	7	1	3	5	3	4
29	4	2	6	4	5	2	4	2	6	1	7	5	6	6	4	0	8	4	1	2
30	4	4	9	8	7	3	4	3	8	2	9	1	5	3	5	9	8	9	2	9
31	6	4	8	0	0	0	4	2	3	8	1	8	4	0	9	5	0	9	0	4
32	3	2	3	8	4	8	8	6	2	9	1	0	1	9	9	3	0	7	3	5
33	6	6	7	2	8	0	0	8	4	0	0	4	6	0	3	2	2	4	6	8
34	8	0	3	1	1	1	1	2	7	0	1	9	1	2	7	1	3	3	5	3
35	5	3	5	7	3	6	3	1	7	2	5	5	1	4	7	1	6	5	6	5
36	9	1	1	9	2	8	3	0	3	6	7	7	4	7	5	9	8	1	8	3
37	9	0	2	9	9	7	4	6	3	6	6	3	7	4	2	7	0	0	1	9
38	8	1	4	6	4	6	8	2	8	9	5	5	2	9	6	2	5	3	0	3
39	4	1	1	9	7	0	7	2	9	0	9	7	0	4	6	2	3	1	0	9
40	9	9	2	7	1	3	2	9	0	3	9	0	7	5	6	7	1	7	8	7

This page intentionally left blank

ID	Travel to School	Favorite Season	Allergies	Favorite School Subject	Favorite Music	Superpower
1	Car	Spring	Yes	English	Pop	Freeze time
2	Car	Summer	Yes	Music	Pop	Telepathy
3	Car	Summer	No	Science	Pop	Fly
4	Walk	Fall	No	Computers and technology	Pop	Invisibility
5	Car	Summer	No	Art	Country	Telepathy
6	Car	Summer	No	Physical education	Rap/Hip-hop	Freeze time
7	Car	Spring	No	Physical education	Pop	Telepathy
8	Car	Winter	No	Art	Other	Fly
9	Car	Summer	No	Physical education	Pop	Fly
10	Car	Spring	No	Mathematics and statistics	Pop	Telepathy
11	Car	Summer	Yes	History	Rap/Hip-hop	Invisibility
12	Car	Spring	No	Art	Rap/Hip-hop	Freeze time
13	Bus	Winter	No	Computers and technology	Rap/Hip-hop	Fly
14	Car	Winter	Yes	Social studies	Rap/Hip-hop	Fly
15	Car	Summer	No	Art	Pop	Freeze time
16	Car	Fall	No	Mathematics and statistics	Pop	Fly
17	Bus	Winter	No	Science	Rap/Hip-hop	Freeze time
18	Car	Spring	Yes	Art	Pop	Telepathy
19	Car	Fall	Yes	Science	Pop	Telepathy
20	Car	Summer	Yes	Physical education	Rap/Hip-hop	Invisibility
21	Car	Spring	Yes	Science	Pop	Invisibility
22	Car	Winter	Yes	Mathematics and statistics	Country	Invisibility
23	Car	Summer	Yes	Art	Pop	Invisibility
24	Bus	Winter	Yes	Other	Pop	Telepathy
25	Bus	Summer	Yes	Science	Other	Fly
26	Car	Summer	No	Science	Pop	Fly
27	Car	Summer	Yes	Music	Pop	Telepathy
28	Car	Summer	No	Physical education	Country	Super strength
29	Car	Fall	Yes	Mathematics and statistics	Country	Telepathy
30	Car	Summer	Yes	Physical education	Rap/Hip-hop	Telepathy
31	Boat	Winter	No	Computers and technology	Gospel	Invisibility
32	Car	Spring	No	Physical education	Pop	Fly
33	Car	Spring	No	Physical education	Pop	Fly
34	Car	Summer	No	Mathematics and statistics	Classical	Fly
35	Car	Fall	Yes	Science	Jazz	Telepathy
36	Car	Spring	No	Science	Rap/Hip-hop	Telepathy
37	Car	Summer	No	Music	Country	Telepathy
38	Bus	Winter	No	Mathematics and statistics	Pop	Fly
39	Car	Spring	No	Art	Classical	Freeze time
40	Car	Winter	Yes	Art	Pop	Fly
41	Walk	Summer	Yes	Physical education	Rap/Hip-hop	Fly
42	Bus	Winter	Yes	Physical education	Gospel	Invisibility

Lesson 20:   Estimating a Population Proportion

S.143

43	Bus	Summer	No	Art	Other	Invisibility
44	Car	Summer	Yes	Computers and technology	Other	Freeze time
45	Car	Fall	Yes	Science	Pop	Fly
46	Car	Summer	Yes	Music	Rap/Hip-hop	Fly
47	Car	Spring	No	Science	Rap/Hip-hop	Invisibility
48	Bus	Spring	No	Music	Pop	Telepathy
49	Car	Summer	Yes	Social studies	Techno/Electronic	Telepathy
50	Car	Summer	Yes	Physical education	Pop	Telepathy
51	Car	Spring	Yes	Other	Other	Telepathy
52	Car	Summer	No	Art	Pop	Fly
53	Car	Summer	Yes	Other	Pop	Telepathy
54	Car	Summer	Yes	Physical education	Rap/Hip-hop	Invisibility
55	Bus	Summer	Yes	Physical education	Other	Super strength
56	Car	Summer	No	Science	Rap/Hip-hop	Invisibility
57	Car	Winter	No	Languages	Rap/Hip-hop	Super strength
58	Car	Fall	Yes	English	Pop	Fly
59	Car	Winter	No	Science	Pop	Telepathy
60	Car	Summer	No	Art	Pop	Invisibility
61	Car	Summer	Yes	Other	Pop	Freeze time
62	Bus	Spring	No	Science	Pop	Fly
63	Car	Winter	Yes	Mathematics and statistics	Other	Freeze time
64	Car	Summer	No	Social studies	Classical	Fly
65	Car	Winter	Yes	Science	Pop	Telepathy
66	Car	Winter	No	Science	Rock and roll	Fly
67	Car	Summer	No	Mathematics and statistics	Rap/Hip-hop	Super strength
68	Car	Fall	No	Music	Rock and roll	Super strength
69	Car	Spring	No	Other	Other	Invisibility
70	Car	Summer	Yes	Mathematics and statistics	Rap/Hip-hop	Telepathy
71	Car	Winter	No	Art	Other	Fly
72	Car	Spring	Yes	Mathematics and statistics	Pop	Telepathy
73	Car	Winter	Yes	Computers and technology	Techno/Electronic	Telepathy
74	Walk	Winter	No	Physical education	Techno/Electronic	Fly
75	Walk	Summer	No	History	Rock and roll	Fly
76	Skateboard/Scooter/Rollerblade	Winter	Yes	Computers and technology	Techno/Electronic	Freeze time
77	Car	Spring	Yes	Science	Other	Telepathy
78	Car	Summer	No	Music	Rap/Hip-hop	Invisibility
79	Car	Summer	No	Social studies	Pop	Invisibility
80	Car	Summer	No	Other	Rap/Hip-hop	Telepathy
81	Walk	Spring	Yes	History	Rap/Hip-hop	Invisibility
82	Car	Summer	No	Art	Pop	Invisibility

Lesson 20:   Estimating a Population Proportion

83	Walk	Spring	No	Languages	Jazz	Super strength
84	Car	Fall	No	History	Jazz	Invisibility
85	Car	Summer	No	Physical education	Rap/Hip-hop	Freeze time
86	Car	Spring	No	Mathematics and statistics	Pop	Freeze time
87	Bus	Spring	Yes	Art	Pop	Telepathy
88	Car	Winter	No	Mathematics and statistics	Other	Invisibility
89	Car	Summer	Yes	Physical education	Country	Telepathy
90	Bus	Summer	No	Computers and technology	Other	Fly
91	Car	Winter	No	History	Pop	Telepathy
92	Walk	Winter	No	Science	Classical	Telepathy
93	Bicycle	Summer	No	Physical education	Pop	Invisibility
94	Car	Summer	No	English	Pop	Telepathy
95	Car	Summer	Yes	Physical education	Pop	Fly
96	Car	Winter	No	Science	Other	Freeze time
97	Car	Winter	No	Other	Rap/Hip-hop	Super strength
98	Car	Summer	Yes	Physical education	Rap/Hip-hop	Freeze time
99	Car	Spring	No	Music	Classical	Telepathy
100	Car	Spring	Yes	Science	Gospel	Telepathy
101	Car	Summer	Yes	History	Pop	Super strength
102	Car	Winter	Yes	English	Country	Freeze time
103	Car	Spring	No	Computers and technology	Other	Telepathy
104	Car	Winter	No	History	Other	Invisibility
105	Car	Fall	No	Music	Pop	Telepathy
106	Car	Fall	No	Science	Pop	Telepathy
107	Car	Winter	No	Art	Heavy metal	Fly
108	Car	Spring	Yes	Science	Rock and roll	Fly
109	Car	Fall	Yes	Music	Other	Fly
110	Car	Summer	Yes	Social studies	Techno/Electronic	Telepathy
111	Car	Spring	No	Physical education	Pop	Fly
112	Car	Summer	No	Physical education	Pop	Fly
113	Car	Summer	Yes	Social studies	Pop	Freeze time
114	Car	Summer	Yes	Computers and technology	Gospel	Freeze time
115	Car	Winter	Yes	Other	Rap/Hip-hop	Telepathy
116	Car	Summer	Yes	Science	Country	Telepathy
117	Car	Fall		Music	Country	Fly
118	Walk	Summer	No	History	Pop	Telepathy
119	Car	Spring	Yes	Art	Pop	Freeze time
120	Car	Fall	Yes	Physical education	Rap/Hip-hop	Fly
121	Car	Spring	No	Music	Rock and roll	Telepathy
122	Car	Fall	No	Art	Pop	Invisibility
123	Car	Summer	Yes	Physical education	Rap/Hip-hop	Fly
124	Walk	Summer	No	Computers and technology	Pop	Telepathy
125	Car	Fall	No	Art	Pop	Fly

Lesson 20:   Estimating a Population Proportion

126	Bicycle	Spring	No	Science	Pop	Invisibility
127	Car	Summer	No	Social studies	Gospel	Fly
128	Bicycle	Winter	No	Social studies	Rap/Hip-hop	Fly
129	Car	Summer	Yes	Mathematics and statistics	Pop	Invisibility
130	Car	Fall	Yes	Mathematics and statistics	Country	Telepathy
131	Car	Winter	Yes	Music	Gospel	Super strength
132	Rail (Train/Tram/ Subway)	Fall	Yes	Art	Other	Fly
133	Walk	Summer	No	Social studies	Pop	Invisibility
134	Car	Summer	Yes	Music	Pop	Freeze time
135	Car	Winter	No	Mathematics and statistics	Pop	Telepathy
136	Car	Fall	Yes	Music	Pop	Telepathy
137	Car	Summer	Yes	Computers and technology	Other	Freeze time
138	Car	Summer	Yes	Physical education	Pop	Telepathy
139	Car	Summer	Yes	Social studies	Other	Telepathy
140	Car	Spring	Yes	Physical education	Other	Freeze time
141	Car	Fall	Yes	Science	Country	Telepathy
142	Car	Spring	Yes	Science	Pop	Invisibility
143	Car	Summer	No	Other	Rap/Hip-hop	Freeze time
144	Car	Summer	No	Other	Other	Fly
145	Car	Summer	No	Languages	Pop	Freeze time
146	Car	Summer	Yes	Physical education	Pop	Telepathy
147	Bus	Winter	No	History	Country	Invisibility
148	Car	Spring	No	Computers and technology	Other	Telepathy
149	Bus	Winter	Yes	Science	Pop	Invisibility
150	Car	Summer	No	Social studies	Rap/Hip-hop	Invisibility
151	Car	Summer	No	Physical education	Pop	Invisibility
152	Car	Summer	Yes	Physical education	Pop	Super strength
153	Car	Summer	No	Mathematics and statistics	Pop	Fly
154	Car	Summer	No	Art	Rap/Hip-hop	Freeze time
155	Car	Winter	Yes	Other	Classical	Freeze time
156	Car	Summer	Yes	Computers and technology	Other	Telepathy
157	Car	Spring	No	Other	Pop	Freeze time
158	Car	Winter	Yes	Music	Country	Fly
159	Car	Winter	No	History	Jazz	Invisibility
160	Car	Spring	Yes	History	Pop	Fly
161	Car	Winter	Yes	Mathematics and statistics	Other	Telepathy
162	Car	Fall	No	Science	Country	Invisibility
163	Car	Winter	No	Science	Other	Fly
164	Car	Summer	No	Science	Pop	Fly
165	Skateboard/ Scooter/ Rollerblade	Spring	Yes	Social studies	Other	Freeze time
166	Car	Winter	Yes	Art	Rap/Hip-hop	Fly

167	Car	Summer	Yes	Other	Pop	Freeze time
168	Car	Summer	No	English	Pop	Telepathy
169	Car	Summer	No	Other	Pop	Invisibility
170	Car	Summer	Yes	Physical education	Techno/Electronic	Freeze time
171	Car	Summer	No	Art	Pop	Telepathy
172	Car	Summer	No	Physical education	Rap/Hip-hop	Freeze time
173	Car	Winter	Yes	Mathematics and statistics	Other	Invisibility
174	Bus	Summer	Yes	Music	Pop	Freeze time
175	Car	Winter	No	Art	Pop	Fly
176	Car	Fall	No	Science	Rap/Hip-hop	Fly
177	Car	Winter	Yes	Social studies	Pop	Telepathy
178	Car	Fall	No	Art	Other	Fly
179	Bus	Spring	No	Physical education	Country	Fly
180	Car	Winter	No	Music	Other	Telepathy
181	Bus	Summer	No	Computers and technology	Rap/Hip-hop	Freeze time
182	Car	Summer	Yes	Physical education	Rap/Hip-hop	Invisibility
183	Car	Summer	Yes	Music	Other	Telepathy
184	Car	Spring	No	Science	Rap/Hip-hop	Invisibility
185	Rail (Train/Tram/Subway)	Summer	No	Physical education	Other	Freeze time
186	Car	Summer	Yes	Mathematics and statistics	Rap/Hip-hop	Fly
187	Bus	Winter	Yes	Mathematics and statistics	Other	Super strength
188	Car	Summer	No	Mathematics and statistics	Other	Freeze time
189	Rail (Train/Tram/Subway)	Fall	Yes	Music	Jazz	Fly
190	Car	Summer	Yes	Science	Pop	Super strength
191	Car	Summer	Yes	Science	Techno/Electronic	Freeze time
192	Car	Spring	Yes	Physical education	Rap/Hip-hop	Freeze time
193	Car	Summer	Yes	Physical education	Rap/Hip-hop	Freeze time
194	Car	Winter	No	Physical education	Rap/Hip-hop	Telepathy
195	Car	Winter	No	Music	Jazz	Freeze time
196	Walk	Summer	Yes	History	Country	Freeze time
197	Car	Spring	No	History	Rap/Hip-hop	Freeze time
198	Car	Fall	Yes	Other	Pop	Freeze time
199	Car	Spring	Yes	Science	Other	Freeze time
200	Bicycle	Winter	Yes	Other	Rap/Hip-hop	Freeze time

Lesson 20:   Estimating a Population Proportion

This page intentionally left blank

Lesson 21: Why Worry About Sampling Variability?

Classwork

There are three bags, Bag A, Bag B, and Bag C, with 100 numbers in each bag. You and your classmates will investigate the population mean (the mean of all 100 numbers) in each bag. Each set of numbers has the same range. However, the population means of each set may or may not be the same. We will see who can uncover the mystery of the bags!

Exercises

1. To begin your investigation, start by selecting a random sample of ten numbers from Bag A. Remember to mix the numbers in the bag first. Then, select one number from the bag. Do not put it back into the bag. Write the number in the chart below. Continue selecting one number at a time until you have selected ten numbers. Mix up the numbers in the bag between each selection.

Selection	1	2	3	4	5	6	7	8	9	10
Bag A										

    a. Create a dot plot of your sample of ten numbers. Use a dot to represent each number in the sample.

    b. Do you think the mean of all the numbers in Bag A might be 10? Why or why not?

    c. Based on the dot plot, what would you estimate the mean of the numbers in Bag A to be? How did you make your estimate?

    d. Do you think your sample mean will be close to the population mean? Why or why not?

   e.   Is your sample mean the same as your neighbors' sample means? Why or why not?

2.   Repeat the process by selecting a random sample of ten numbers from Bag B.

Selection	1	2	3	4	5	6	7	8	9	10
Bag B										

   a.   Create a dot plot of your sample of ten numbers. Use a dot to represent each of the numbers in the sample.

   b.   Based on your dot plot, do you think the mean of the numbers in Bag B is the same or different from the mean of the numbers in Bag A? Explain your thinking.

3.   Repeat the process once more by selecting a random sample of ten numbers from Bag C.

Selection	1	2	3	4	5	6	7	8	9	10
Bag C										

   a.   Create a dot plot of your sample of ten numbers. Use a dot to represent each of the numbers in the sample.

   b.   Based on your dot plot, do you think the mean of the numbers in Bag C is the same as or different from the mean of the numbers in Bag A? Explain your thinking.

4. Are your dot plots of the three bags the same as the dot plots of other students in your class? Why or why not?

5. Calculate the mean of the numbers for each of the samples from Bag A, Bag B, and Bag C.

	Mean of the Sample of Numbers
Bag A	
Bag B	
Bag C	

a. Are the sample means you calculated the same as the sample means of other members of your class? Why or why not?

b. How do your sample means for Bag A and for Bag B compare?

c. Calculate the difference of the sample mean for Bag A minus the sample mean for Bag B (Mean A – Mean B). Based on this difference, can you be sure which bag has the larger population mean? Why or why not?

6. Based on the class dot plots of the sample means, do you think the mean of the numbers in Bag A and the mean of the numbers in Bag B are different? Do you think the mean of the numbers in Bag A and the mean of the numbers in Bag C are different? Explain your answers.

7. Based on the difference between the sample mean of Bag A and the sample mean of Bag B (Mean A − Mean B) that you calculated in Exercise 5, do you think that the two populations (Bags A and B) have different means, or do you think that the two population means might be the same?

8. Based on this difference, can you be sure which bag has the larger population mean? Why or why not?

9. Is your difference in sample means the same as your neighbors' differences? Why or why not?

10. Plot your difference of the means (Mean A − Mean B) on a class dot plot. Describe the distribution of differences plotted on the graph. Remember to discuss center and spread.

11. Why are the differences in the sample means of Bag A and Bag B not always 0?

12. Does the class dot plot contain differences that were relatively far away from 0? If yes, why do you think this happened?

13. Suppose you will take a sample from a new bag. How big would the difference in the sample mean for Bag A and the sample mean for the new bag (Mean A – Mean new) have to be before you would be convinced that the population mean for the new bag is different from the population mean of Bag A? Use the class dot plot of the differences in sample means for Bags A and B (which have equal population means) to help you answer this question.

The differences in the class dot plot occur because of sampling variability—the chance variability from one sample to another. In Exercise 13, you were asked about how great the difference in sample means would need to be before you have convincing evidence that one population mean is larger than another population mean. A *meaningful* difference between two sample means is one that is unlikely to have occurred by chance if the population means are equal. In other words, the difference is one that is greater than would have been expected just due to sampling variability.

14. Calculate the sample mean of Bag A minus the sample mean of Bag C (Mean A – Mean C).

15. Plot your difference (Mean A – Mean C) on a class dot plot.

16. How do the centers of the class dot plots for Mean A – Mean B and Mean A – Mean C compare?

17. Each bag has a population mean that is either 10.5 or 14.5. State what you think the population mean is for each bag. Explain your choice for each bag.

> **Lesson Summary**
>
> - Remember to think about sampling variability—the chance variability from sample to sample.
> - Beware of making decisions based just on the fact that two sample means are not equal.
> - Consider the distribution of the difference in sample means when making a decision.

Problem Set

Below are three dot plots. Each dot plot represents the differences in sample means for random samples selected from two populations (Bag A and Bag B). For each distribution, the differences were found by subtracting the sample means of Bag B from the sample means of Bag A (sample mean A — sample mean B).

1. Does the graph below indicate that the population mean of Bag A is larger than the population mean of Bag B? Why or why not?

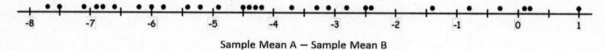

Sample Mean A — Sample Mean B

2. Use the graph above to estimate the difference in the population means (Mean A — Mean B).

3. Does the graph below indicate that the population mean of Bag A is larger than the population mean of Bag B? Why or why not?

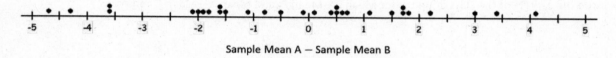

Sample Mean A — Sample Mean B

4. Does the graph below indicate that the population mean of Bag A is larger than the population mean of Bag B? Why or why not?

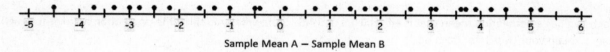

Sample Mean A — Sample Mean B

5. In the above graph, how many differences are greater than 0? How many differences are less than 0? What might this tell you?

6. In Problem 4, the population mean for Bag A is really larger than the population mean for Bag B. Why is it possible to still get so many negative differences in the graph?

Lesson 21:   Why Worry About Sampling Variability?

Lesson 22: Using Sample Data to Compare the Means of Two or More Populations

Classwork

In previous lessons, you worked with one population. Many statistical questions involve comparing two populations.
For example:

- On average, do boys and girls differ on quantitative reasoning?
- Do students learn basic arithmetic skills better with or without calculators?
- Which of two medications is more effective in treating migraine headaches?
- Does one type of car get better mileage per gallon of gasoline than another type?
- Does one type of fabric decay in landfills faster than another type?
- Do people with diabetes heal more slowly than people who do not have diabetes?

In this lesson, you will begin to explore how big of a difference there needs to be in sample means in order for the
difference to be considered meaningful. The next lesson will extend that understanding to making informal inferences
about population differences.

Examples 1–3

Tamika's mathematics project is to see whether boys or girls are faster in solving a KenKen-type puzzle. She creates a
puzzle and records the following times that it took to solve the puzzle (in seconds) for a random sample of 10 boys from
her school and a random sample of 11 girls from her school:

												Mean	MAD
Boys	39	38	27	36	40	27	43	36	34	33			35.3	4.04
Girls	41	41	33	42	47	38	41	36	36	32	46		39.4	3.96

1. On the same scale, draw dot plots for the boys' data and for the girls' data. Comment on the amount of overlap
    between the two dot plots. How are the dot plots the same, and how are they different?

---

2. Compare the variability in the two data sets using the MAD (mean absolute deviation). Is the variability in each sample about the same? Interpret the MAD in the context of the problem.

3. In the previous lesson, you learned that a difference between two sample means is considered to be meaningful if the difference is more than what you would expect to see just based on sampling variability. The difference in the sample means of the boys' times and the girls' times is 4.1 seconds (39.4 seconds − 35.3 seconds). This difference is approximately 1 MAD.

   a. If 4 *sec.* is used to approximate the values of 1 MAD for both boys and for girls, what is the interval of times that are within 1 MAD of the sample mean for boys?

   b. Of the 10 sample means for boys, how many of them are within that interval?

   c. Of the 11 sample means for girls, how many of them are within the interval you calculated in part (a)?

   d. Based on the dot plots, do you think that the difference between the two sample means is a meaningful difference? That is, are you convinced that the mean time for all girls at the school (not just this sample of girls) is different from the mean time for all boys at the school? Explain your choice based on the dot plots.

Examples 4–7

How good are you at estimating a minute? Work in pairs. Flip a coin to determine which person in the pair will go first. One of you puts your head down and raises your hand. When your partner says "Start," keep your head down and your hand raised. When you think a minute is up, put your hand down. Your partner will record how much time has passed. Note that the room needs to be quiet. Switch roles, except this time you talk with your partner during the period when the person with his head down is indicating when he thinks a minute is up. Note that the room will not be quiet.

Group	Estimate for a Minute													
Quiet														
Talking														

Use your class data to complete the following.

4. Calculate the mean minute time for each group. Then, find the difference between the *quiet* mean and the *talking* mean.

5. On the same scale, draw dot plots of the two data distributions, and discuss the similarities and differences in the two distributions.

6. Calculate the mean absolute deviation (MAD) for each data set. Based on the MADs, compare the variability in each sample. Is the variability about the same? Interpret the MADs in the context of the problem.

7. Based on your calculations, is the difference in mean time estimates meaningful? Part of your reasoning should involve the number of MADs that separate the two sample means. Note that if the MADs differ, use the larger one in determining how many MADs separate the two means.

©2015 Great Minds. eureka-math.org
G7-M5-SE-B3-1.3.1-01.2016

Lesson Summary

Variability is a natural occurrence in data distributions. Two data distributions can be compared by describing how far apart their sample means are. The amount of separation can be measured in terms of how many MADs separate the means. (Note that if the two sample MADs differ, the larger of the two is used to make this calculation.)

Problem Set

1. A school is trying to decide which reading program to purchase.

   a. How many MADs separate the mean reading comprehension score for a standard program (mean = 67.8, MAD = 4.6, $n = 24$) and an activity-based program (mean = 70.3, MAD = 4.5, $n = 27$)?

   b. What recommendation would you make based on this result?

2. Does a football filled with helium go farther than one filled with air? Two identical footballs were used: one filled with helium and one filled with air to the same pressure. Matt was chosen from your team to do the kicking. You did not tell Matt which ball he was kicking. The data (in yards) follow.

Air	25	23	28	29	27	32	24	26	22	27	31	24	33	26	24	28	30
Helium	24	19	25	25	22	24	28	31	22	26	24	23	22	21	21	23	25

	Mean	MAD
Air		
Helium		

   a. Calculate the difference between the sample mean distance for the football filled with air and for the one filled with helium.

   b. On the same scale, draw dot plots of the two distributions, and discuss the variability in each distribution.

   c. Calculate the MAD for each distribution. Based on the MADs, compare the variability in each distribution. Is the variability about the same? Interpret the MADs in the context of the problem.

   d. Based on your calculations, is the difference in mean distance meaningful? Part of your reasoning should involve the number of MADs that separate the sample means. Note that if the MADs differ, use the larger one in determining how many MADs separate the two means.

---

3. Suppose that your classmates were debating about whether going to college is really worth it. Based on the following data of annual salaries (rounded to the nearest thousands of dollars) for college graduates and high school graduates with no college experience, does it appear that going to college is indeed worth the effort? The data are from people in their second year of employment.

College Grad	41	67	53	48	45	60	59	55	52	52	50	59	44	49	52
High School Grad	23	33	36	29	25	43	42	38	27	25	33	41	29	33	35

a. Calculate the difference between the sample mean salary for college graduates and for high school graduates.

b. On the same scale, draw dot plots of the two distributions, and discuss the variability in each distribution.

c. Calculate the MAD for each distribution. Based on the MADs, compare the variability in each distribution. Is the variability about the same? Interpret the MADs in the context of the problem.

d. Based on your calculations, is going to college worth the effort? Part of your reasoning should involve the number of MADs that separate the sample means.

Lesson 23: Using Sample Data to Compare the Means of Two or More Populations

Classwork

In the previous lesson, you described how far apart the means of two data sets are in terms of the MAD (mean absolute deviation), a measure of variability. In this lesson, you will extend that idea to informally determine when two sample means computed from random samples are far enough apart from each other to imply that the population means also differ in a *meaningful* way. Recall that a *meaningful* difference between two means is a difference that is greater than would have been expected just due to sampling variability.

Example 1: Texting

With texting becoming so popular, Linda wanted to determine if middle school students memorize *real* words more or less easily than *fake* words. For example, real words are *food, car, study, swim,* whereas fake words are *stk, fonw, cqur, ttnsp.* She randomly selected 28 students from all middle school students in her district and gave half of them a list of 20 real words and the other half a list of 20 fake words.

  a.   How do you think Linda might have randomly selected 28 students from all middle school students in her district?

  b.   Why do you think Linda selected the students for her study randomly? Explain.

c.   She gave the selected students one minute to memorize their lists, after which they were to turn the lists over and, after two minutes, write down all the words that they could remember. Afterward, they calculated the number of correct words that they were able to write down. Do you think a penalty should be given for an incorrect word written down? Explain your reasoning.

Exercises 1–4

Suppose the data (the number of correct words recalled) she collected were as follows:

For students given the real words list: 8, 11, 12, 8, 4, 7, 9, 12, 12, 9, 14, 11, 5, 10

For students given the fake words list: 3, 5, 4, 4, 4, 7, 11, 9, 7, 7, 1, 3, 3, 7

1.   On the same scale, draw dot plots for the two data sets.

2.   From looking at the dot plots, write a few sentences comparing the distribution of the number of correctly recalled real words with the distribution of the number of correctly recalled fake words. In particular, comment on which type of word, if either, that students recall better. Explain.

3. Linda made the following calculations for the two data sets:

	Mean	MAD
Real Words Recalled	9.43	2.29
Fake Words Recalled	5.36	2.27

In the previous lesson, you calculated the number of MADs that separated two sample means. You used the larger MAD to make this calculation if the two MADs were not the same. How many MADs separate the mean number of real words recalled and the mean number of fake words recalled for the students in the study?

4. In the last lesson, our work suggested that if the number of MADs that separate the two sample means is 2 or more, then it is reasonable to conclude that not only do the means differ in the samples but that the means differ in the populations as well. If the number of MADs is less than 2, then you can conclude that the difference in the sample means might just be sampling variability and that there may not be a meaningful difference in the population means. Using these criteria, what can Linda conclude about the difference in population means based on the sample data that she collected? Be sure to express your conclusion in the context of this problem.

Example 2

Ken, an eighth-grade student, was interested in doing a statistics study involving sixth-grade and eleventh-grade students in his school district. He conducted a survey on four numerical variables and two categorical variables (grade level and gender). His Excel population database for the 265 sixth graders and 175 eleventh graders in his district has the following description:

Column	Name	Description
1	ID	ID numbers are from 1 through 440. 1–128 sixth-grade females 129–265 sixth-grade males 266–363 eleventh-grade females 364–440 eleventh-grade males
2	Texting	Number of minutes per day texting (whole number)
3	ReacTime	Time in seconds to respond to a computer screen stimulus (two decimal places)
4	Homework	Total number of hours per week spent on doing homework (one decimal place)
5	Sleep	Number of hours per night sleeping (one decimal place)

©2015 Great Minds. eureka-math.org
G7-M5-SE-B3-1.3.1-01.2016

a. Ken decides to base his study on a random sample of 20 sixth graders and a random sample of 20 eleventh graders. The sixth graders have IDs 1–265, and the eleventh graders are numbered 266–440. Advise him on how to randomly sample 20 sixth graders and 20 eleventh graders from his data file.

Suppose that from a random number generator:

The random ID numbers for Ken's 20 sixth graders:
231 15 19 206 86 183 233 253 142 36 195 139 75 210 56 40 66 114 127 9

The random ID numbers for his 20 eleventh graders:
391 319 343 426 307 360 289 328 390 350 279 283 302 287 269 332 414 267 428 280

b. For each set, find the homework hours data from the population database that correspond to these randomly selected ID numbers.

c. On the same scale, draw dot plots for the two sample data sets.

d. From looking at the dot plots, list some observations comparing the number of hours per week that sixth graders spend on doing homework and the number of hours per week that eleventh graders spend on doing homework.

e. Calculate the mean and MAD for each of the data sets. How many MADs separate the two sample means? (Use the larger MAD to make this calculation if the sample MADs are not the same.)

	Mean (hours)	MAD (hours)
Sixth Grade		
Eleventh Grade		

f. Ken recalled Linda suggesting that if the number of MADs is greater than or equal to 2, then it would be reasonable to think that the population of all sixth-grade students in his district and the population of all eleventh-grade students in his district have different means. What should Ken conclude based on his homework study?

EUREKA
MATH

Problem Set

1. Based on Ken's population database, compare the amount of sleep that sixth-grade females get on average to the amount of sleep that eleventh-grade females get on average.

   Find the data for 15 sixth-grade females based on the following random ID numbers:
   65 1 67 101 106 87 85 95 120 4 64 74 102 31 128

   Find the data for 15 eleventh-grade females based on the following random ID numbers:
   348 313 297 351 294 343 275 354 311 328 274 305 288 267 301

2. On the same scale, draw dot plots for the two sample data sets.

3. Looking at the dot plots, list some observations comparing the number of hours per week that sixth graders spend on doing homework and the number of hours per week that eleventh graders spend on doing homework.

4. Calculate the mean and MAD for each of the data sets. How many MADs separate the two sample means? (Use the larger MAD to make this calculation if the sample MADs are not the same.)

	Mean (hours)	MAD (hours)
Sixth-Grade Females		
Eleventh-Grade Females		

5. Recall that if the number of MADs in the difference of two sample means is greater than or equal to 2, then it would be reasonable to think that the population means are different. Using this guideline, what can you say about the average number of hours of sleep per night for all sixth-grade females in the population compared to all eleventh-grade females in the population?

Student Edition

Eureka Math
Grade 7
Module 6

Special thanks go to the Gordon A. Cain Center and to the Department of Mathematics at Louisiana State University for their support in the development of *Eureka Math*.

Lesson 1: Complementary and Supplementary Angles

Classwork

Opening Exercise

As we begin our study of unknown angles, let us review key definitions.

Term	Definition
	Two angles, $\angle AOC$ and $\angle COB$, with a common side $\overrightarrow{OC}$, are _____ angles if C is in the interior of $\angle AOB$.
	When two lines intersect, any two non-adjacent angles formed by those lines are called _____ angles, or _____ _____ angles.
	Two lines are _____ if they intersect in one point, and any of the angles formed by the intersection of the lines is 90°. Two segments or rays are _____ if the lines containing them are _____ lines.

Complete the missing information in the table below. In the *Statement* column, use the illustration to write an equation that demonstrates the angle relationship; use all forms of angle notation in your equations.

Angle Relationship	Abbreviation	Statement	Illustration
Adjacent Angles		The measurements of adjacent angles add.	
Vertical Angles		Vertical angles have equal measures.	

Angles on a Line		If the vertex of a ray lies on a line but the ray is not contained in that line, then the sum of measurements of the two angles formed is 180°.	
Angles at a Point		Suppose three or more rays with the same vertex separate the plane into angles with disjointed interiors. Then, the sum of all the measurements of the angles is 360°.	

Angle Relationship	Definition	Diagram
Complementary Angles		
Supplementary Angles		

©2015 Great Minds. eureka-math.org
G7-M6-SE-B3-1.3.1-01.2016

Exercise 1

1. In a complete sentence, describe the relevant angle
   relationships in the diagram. Write an equation for the angle
   relationship shown in the figure and solve for x. Confirm your
   answers by measuring the angle with a protractor.

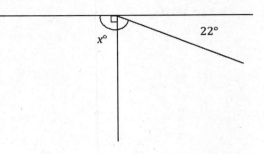

Example 1

The measures of two supplementary angles are in the ratio of $2:3$. Find the measurements of the two angles.

Exercises 2–4

2. In a pair of complementary angles, the measurement of the larger angle is three times that of the smaller angle.
   Find the measurements of the two angles.

3. The measure of a supplement of an angle is 6° more than twice the measure of the angle. Find the measurement of the two angles.

4. The measure of a complement of an angle is 32° more than three times the angle. Find the measurement of the two angles.

Example 2

Two lines meet at a point that is also the vertex of an angle. Set up and solve an appropriate equation for x and y.

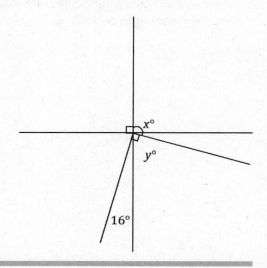

Problem Set

1. Two lines meet at a point that is also the endpoint of a ray. Set up and solve the appropriate equations to determine x and y.

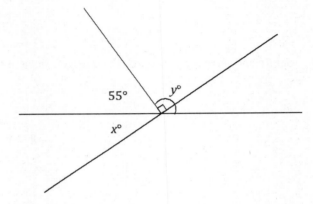

2. Two lines meet at a point that is also the vertex of an angle. Set up and solve the appropriate equations to determine x and y.

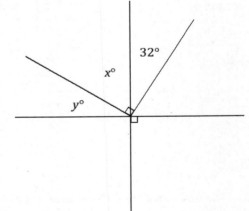

3. Two lines meet at a point that is also the vertex of an angle. Set up and solve an appropriate equation for x and y.

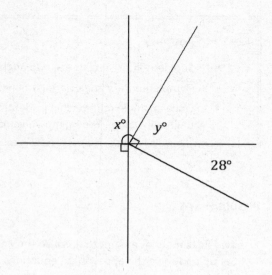

4. Set up and solve the appropriate equations for s and t.

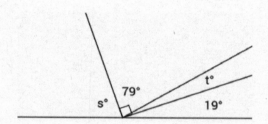

5. Two lines meet at a point that is also the endpoint of two rays. Set up and solve the appropriate equations for m and n.

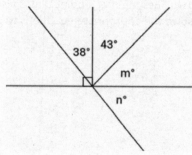

6. The supplement of the measurement of an angle is 16° less than three times the angle. Find the measurement of the angle and its supplement.

7. The measurement of the complement of an angle exceeds the measure of the angle by 25%. Find the measurement of the angle and its complement.

8. The ratio of the measurement of an angle to its complement is $1:2$. Find the measurement of the angle and its complement.

9. The ratio of the measurement of an angle to its supplement is $3:5$. Find the measurement of the angle and its supplement.

10. Let x represent the measurement of an acute angle in degrees. The ratio of the complement of x to the supplement of x is $2:5$. Guess and check to determine the value of x. Explain why your answer is correct.

This page intentionally left blank

Lesson 2: Solving for Unknown Angles Using Equations

Classwork

Opening Exercise

Two lines meet at a point. In a complete sentence, describe the relevant angle relationships in the diagram. Find the values of r, s, and t.

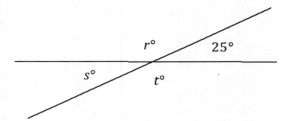

Example 1

Two lines meet at a point that is also the endpoint of a ray. In a complete sentence, describe the relevant angle relationships in the diagram. Set up and solve an equation to find the value of p and r.

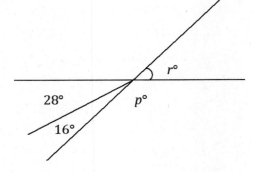

EUREKA
MATH™

Exercise 1

Three lines meet at a point. In a complete sentence, describe the relevant angle relationship in the diagram. Set up and solve an equation to find the value of a.

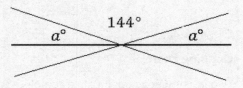

Example 2

Three lines meet at a point. In a complete sentence, describe the relevant angle relationships in the diagram. Set up and solve an equation to find the value of z.

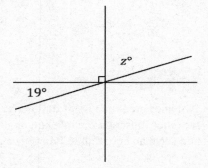

EUREKA
MATH™

Exercise 2

Three lines meet at a point; $\angle AOF = 144°$. In a complete sentence, describe the relevant angle relationships in the diagram. Set up and solve an equation to determine the value of c.

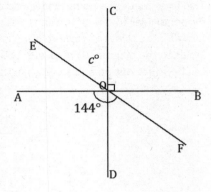

Example 3

Two lines meet at a point that is also the endpoint of a ray. The ray is perpendicular to one of the lines as shown. In a complete sentence, describe the relevant angle relationships in the diagram. Set up and solve an equation to find the value of t.

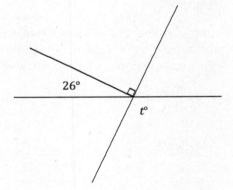

Exercise 3

Two lines meet at a point that is also the endpoint of a ray. The ray is perpendicular to one of the lines as shown. In a complete sentence, describe the relevant angle relationships in the diagram. You may add labels to the diagram to help with your description of the angle relationship. Set up and solve an equation to find the value of v.

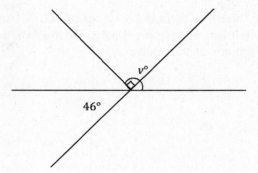

Example 4

Three lines meet at a point. In a complete sentence, describe the relevant angle relationships in the diagram. Set up and solve an equation to find the value of x. Is your answer reasonable? Explain how you know.

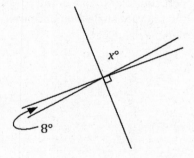

EUREKA
MATH™

Exercise 4

Two lines meet at a point that is also the endpoint of two rays. In a complete sentence, describe the relevant angle relationships in the diagram. Set up and solve an equation to find the value of x. Find the measurements of $\angle AOB$ and $\angle BOC$.

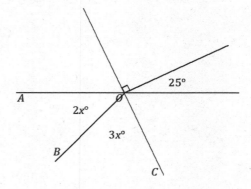

Exercise 5

a. In a complete sentence, describe the relevant angle relationships in the diagram. Set up and solve an equation to find the value of x. Find the measurements of $\angle AOB$ and $\angle BOC$.

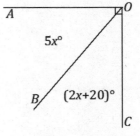

b. Katrina was solving the problem above and wrote the equation $7x + 20 = 90$. Then, she rewrote this as $7x + 20 = 70 + 20$. Why did she rewrite the equation in this way? How does this help her to find the value of x?

Lesson Summary

- To solve an unknown angle problem, identify the angle relationship(s) first to set up an equation that will yield the unknown value.

- Angles on a line and supplementary angles are not the same relationship. *Supplementary* angles are two angles whose angle measures sum to 180° whereas *angles on a line* are two or more adjacent angles whose angle measures sum to 180°.

Problem Set

1. Two lines meet at a point that is also the endpoint of a ray.
    Set up and solve an equation to find the value of c.

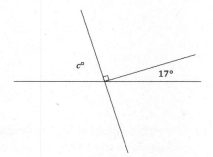

2. Two lines meet at a point that is also the endpoint of a ray. Set up and solve an equation to find the value of a. Explain why your answer is reasonable.

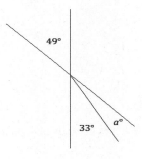

3. Two lines meet at a point that is also the endpoint of a ray. Set up and solve an equation to find the value of w.

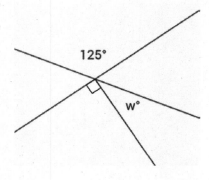

4. Two lines meet at a point that is also the vertex of an angle. Set up and solve an equation to find the value of m.

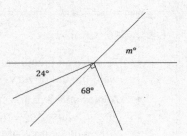

5. Three lines meet at a point. Set up and solve an equation to find the value of r.

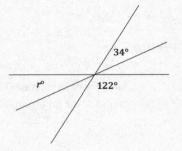

6. Three lines meet at a point that is also the endpoint of a ray. Set up and solve an equation to find the value of each variable in the diagram.

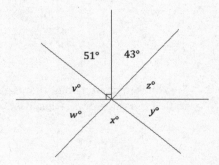

7. Set up and solve an equation to find the value of x. Find the measurement of $\angle AOB$ and of $\angle BOC$.

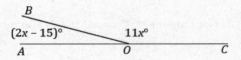

Lesson 2:   Solving for Unknown Angles Using Equations

EUREKA
MATH™

8. Set up and solve an equation to find the value of x. Find the measurement of $\angle AOB$ and of $\angle BOC$.

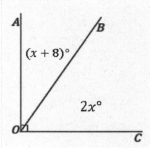

9. Set up and solve an equation to find the value of x. Find the measurement of $\angle AOB$ and of $\angle BOC$.

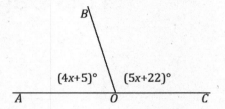

10. Write a verbal problem that models the following diagram. Then, solve for the two angles.

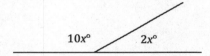

This page intentionally left blank

Lesson 3: Solving for Unknown Angles Using Equations

Classwork

Opening Exercise

Two lines meet at a point that is also the vertex of an angle; the measurement of $\angle AOF$ is 134°. Set up and solve an equation to find the values of x and y. Are your answers reasonable? How do you know?

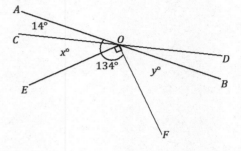

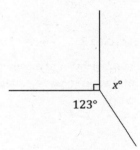

Example 1

Set up and solve an equation to find the value of x.

Exercise 1

Five rays meet at a common endpoint. In a complete sentence, describe the relevant angle relationships in the diagram. Set up and solve an equation to find the value of a.

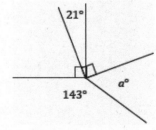

EUREKA MATH™

Example 2

Four rays meet at a common endpoint. In a complete sentence, describe the relevant angle relationships in the diagram. Set up and solve an equation to find the value of x. Find the measurements of $\angle BAC$ and $\angle DAE$.

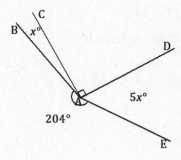

Exercise 2

Four rays meet at a common endpoint. In a complete sentence, describe the relevant angle relationships in the diagram. Set up and solve an equation to find the value of x. Find the measurement of $\angle CAD$.

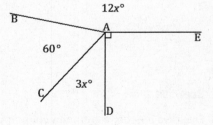

EUREKA
MATH™

Example 3

Two lines meet at a point that is also the endpoint of two rays. In a complete sentence, describe the relevant angle relationships in the diagram. Set up and solve an equation to find the value of x. Find the measurements of $\angle BAC$ and $\angle BAH$.

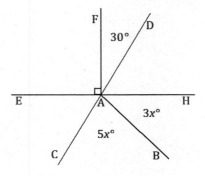

Exercise 3

Lines AB and EF meet at a point which is also the endpoint of two rays. In a complete sentence, describe the relevant angle relationships in the diagram. Set up and solve an equation to find the value of x. Find the measurements of $\angle DHF$ and $\angle AHD$.

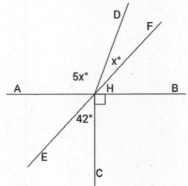

Example 4

Two lines meet at a point. Set up and solve an equation to find the value of x. Find the measurement of one of the vertical angles.

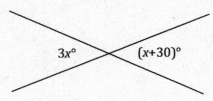

Exercise 4

Set up and solve an equation to find the value of x. Find the measurement of one of the vertical angles.

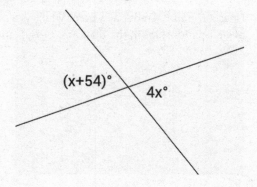

EUREKA
MATH™

Lesson Summary

Steps to Solving for Unknown Angles

- Identify the angle relationship(s).
- Set up an equation that will yield the unknown value.
- Solve the equation for the unknown value.
- Substitute the answer to determine the angle(s).
- Check and verify your answer by measuring the angle with a protractor.

Problem Set

1. Two lines meet at a point. Set up and solve an equation to find the value of x.

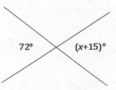

2. Three lines meet at a point. Set up and solve an equation to find the value of a. Is your answer reasonable? Explain how you know.

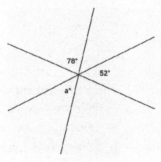

3. Two lines meet at a point that is also the endpoint of two rays. Set up and solve an equation to find the values of a and b.

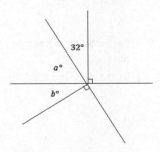

4. Three lines meet at a point that is also the endpoint of a ray. Set up and solve an equation to find the values of x and y.

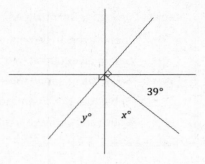

5. Two lines meet at a point. Find the measurement of one of the vertical angles. Is your answer reasonable? Explain how you know.

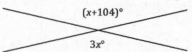

6. Three lines meet at a point that is also the endpoint of a ray. Set up and solve an equation to find the value of y.

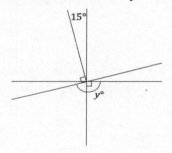

7. Three adjacent angles are at a point. The second angle is 20° more than the first, and the third angle is 20° more than the second angle.
    a. Find the measurements of all three angles.
    b. Compare the expressions you used for the three angles and their combined expression. Explain how they are equal and how they reveal different information about this situation.

8. Four adjacent angles are on a line. The measurements of the four angles are four consecutive even numbers. Determine the measurements of all four angles.

9. Three adjacent angles are at a point. The ratio of the measurement of the second angle to the measurement of the first angle is 4 : 3. The ratio of the measurement of the third angle to the measurement of the second angle is 5 : 4. Determine the measurements of all three angles.

10. Four lines meet at a point. Solve for x and y in the following diagram.

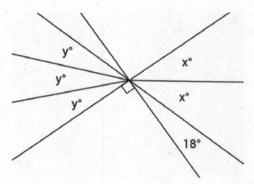

EUREKA
MATH™

This page intentionally left blank

Lesson 4: Solving for Unknown Angles Using Equations

Classwork

Opening Exercise

The complement of an angle is four times the measurement of the angle. Find the measurement of the angle and its complement.

Example 1

Find the measurements of ∠FAE and ∠CAD.

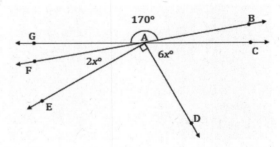

Two lines meet at a point. List the relevant angle relationship in the diagram. Set up and solve an equation to find the value of x. Find the measurement of one of the vertical angles.

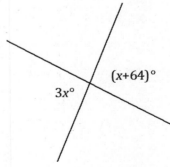

Exercise 1

Set up and solve an equation to find the value of x. List the relevant angle relationship in the diagram. Find the measurement of one of the vertical angles.

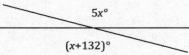

$5x°$

$(x+132)°$

Example 2

Three lines meet at a point. List the relevant angle relationships in the diagram. Set up and solve an equation to find the value of b.

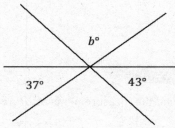

$b°$

$37°$ $43°$

Exercise 2

Two lines meet at a point that is also the endpoint of two rays. List the relevant angle relationships in the diagram. Set up and solve an equation to find the value of b.

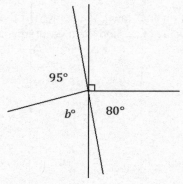

$95°$

$b°$ $80°$

Example 3

The measurement of an angle is $\frac{2}{3}$ the measurement of its supplement. Find the measurements of the angle and its supplement.

Exercise 3

The measurement of an angle is $\frac{1}{4}$ the measurement of its complement. Find the measurements of the two complementary angles.

Example 4

Three lines meet at a point that is also the endpoint of a ray. List the relevant angle relationships in the diagram. Set up and solve an equation to find the value of z.

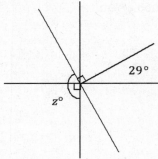

Exercise 4

Two lines meet at a point that is also the vertex of an angle. Set up and solve an equation to find the value of x. Find the measurements of $\angle GAF$ and $\angle BAC$.

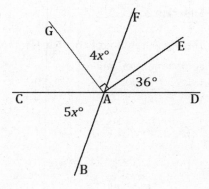

Lesson 4:   Solving for Unknown Angles Using Equations

©2015 Great Minds. eureka-math.org
G7-M6-SE-B3-1.3.1-01.2016

Lesson Summary

Steps to Solving for Unknown Angles

- Identify the angle relationship(s).
- Set up an equation that will yield the unknown value.
- Solve the equation for the unknown value.
- Substitute the answer to determine the measurement of the angle(s).
- Check and verify your answer by measuring the angle with a protractor.

Problem Set

1. Four rays have a common endpoint on a line. Set up and solve an equation to find the value of c.

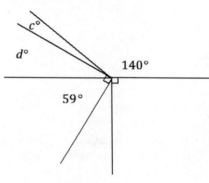

2. Lines BC and EF meet at A. Set up and solve an equation to find the value of x. Find the measurements of $\angle EAH$ and $\angle HAC$.

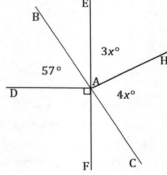

3. Five rays share a common endpoint. Set up and solve an equation to find the value of x. Find the measurements of ∠DAG and ∠GAH.

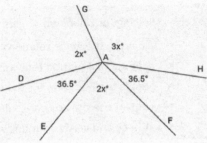

4. Four lines meet at a point which is also the endpoint of three rays. Set up and solve an equation to find the values of x and y.

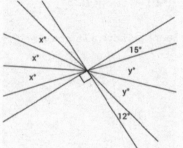

5. Two lines meet at a point that is also the vertex of a right angle. Set up and solve an equation to find the value of x. Find the measurements of ∠CAE and ∠BAG.

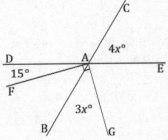

6. Five angles are at a point. The measurement of each angle is one of five consecutive, positive whole numbers.

   a. Determine the measurements of all five angles.

   b. Compare the expressions you used for the five angles and their combined expression. Explain how they are equivalent and how they reveal different information about this situation.

7. Let $x°$ be the measurement of an angle. The ratio of the measurement of the complement of the angle to the measurement of the supplement of the angle is $1:3$. The measurement of the complement of the angle and the measurement of the supplement of the angle have a sum of $180°$. Use a tape diagram to find the measurement of this angle.

EUREKA
MATH™

8. Two lines meet at a point. Set up and solve an equation to find the value of x. Find the measurement of one of the vertical angles.

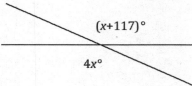

$(x+117)°$

$4x°$

9. The difference between three times the measurement of the complement of an angle and the measurement of the supplement of that angle is 20°. What is the measurement of the angle?

This page intentionally left blank

Lesson 5: Identical Triangles

Classwork

Opening

When studying triangles, it is essential to be able to communicate about the parts of a triangle without any confusion. The following terms are used to identify particular angles or sides:

- between
- adjacent to
- opposite to
- included [side/angle]

Exercises 1–7

Use the figure △ ABC to fill in the following blanks.

1. ∠A is _____ sides $\overline{AB}$ and $\overline{AC}$.

2. ∠B is _____ side $\overline{AB}$ and to side $\overline{BC}$.

3. Side $\overline{AB}$ is _____ ∠C.

4. Side _____ is the included side of ∠B and ∠C.

5. ∠_____ is opposite to side $\overline{AC}$.

6. Side $\overline{AB}$ is between ∠_____ and ∠_____.

7. What is the included angle of sides $\overline{AB}$ and $\overline{BC}$ _____

Now that we know what to call the parts within a triangle, we consider how to discuss two triangles. We need to compare the parts of the triangles in a way that is easy to understand. To establish some alignment between the triangles, we pair up the vertices of the two triangles. We call this a *correspondence*. Specifically, a correspondence between two triangles is a pairing of each vertex of one triangle with one (and only one) vertex of the other triangle. A correspondence provides a systematic way to compare parts of two triangles.

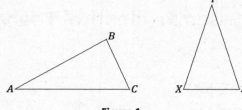

Figure 1

In Figure 1, we can choose to assign a correspondence so that A matches to X, B matches to Y, and C matches to Z. We notate this correspondence with double arrows: $A \leftrightarrow X$, $B \leftrightarrow Y$, and $C \leftrightarrow Z$. This is just one of six possible correspondences between the two triangles. Four of the six correspondences are listed below; find the remaining two correspondences.

A simpler way to indicate the triangle correspondences is to let the order of the vertices define the correspondence (i.e., the first corresponds to the first, the second to the second, and the third to the third). The correspondences above can be written in this manner. Write the remaining two correspondences in this way.

$$\triangle ABC \leftrightarrow \triangle XYZ \qquad \triangle ABC \leftrightarrow \triangle XZY$$

$$\triangle ABC \leftrightarrow \triangle YXZ \qquad \triangle ABC \leftrightarrow \triangle YZX$$

With a correspondence in place, comparisons can be made about corresponding sides and corresponding angles. The following are corresponding vertices, angles, and sides for the triangle correspondence $\triangle ABC \leftrightarrow \triangle YXZ$. Complete the missing correspondences.

Vertices: $\quad A \leftrightarrow Y$	$B \leftrightarrow$	$C \leftrightarrow$
Angles: $\quad \angle A \leftrightarrow \angle Y$	$\angle B \leftrightarrow$	$\angle C \leftrightarrow$
Sides: $\quad \overline{AB} \leftrightarrow \overline{YX}$	$\overline{BC} \leftrightarrow$	$\overline{CA} \leftrightarrow$

EUREKA
MATH™

Example 1

Given the following triangle correspondences, use double arrows to show the correspondence between vertices, angles, and sides.

Triangle Correspondence	$\triangle ABC \leftrightarrow \triangle STR$
Correspondence of Vertices	
Correspondence of Angles	
Correspondence of Sides	

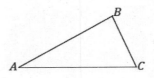

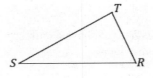

Examine Figure 2. By simply looking, it is impossible to tell the two triangles apart unless they are labeled. They look exactly the same (just as identical twins look the same). One triangle could be picked up and placed on top of the other.

Two triangles are identical if there is a triangle correspondence so that corresponding sides and angles of each triangle are equal in measurement. In Figure 2, there is a correspondence that will match up equal sides and equal angles, $\triangle ABC \leftrightarrow \triangle XYZ$; we can conclude that $\triangle ABC$ is identical to $\triangle XYZ$. This is not to say that we cannot find a correspondence in Figure 2 so that unequal sides and unequal angles are matched up, but there certainly is one correspondence that will match up angles with equal measurements and sides of equal lengths, making the triangles identical.

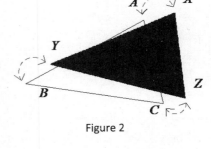

Figure 2

In discussing identical triangles, it is useful to have a way to indicate those sides and angles that are equal. We mark sides with tick marks and angles with arcs if we want to draw attention to them. If two angles or two sides have the same number of marks, it means they are equal.

In this figure, $AC = DE = EF$, and $\angle B = \angle E$.

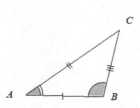

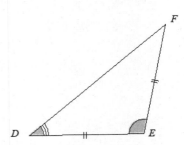

Example 2

Two identical triangles are shown below. Give a triangle correspondence that matches equal sides and equal angles.

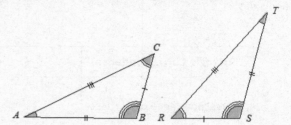

Exercise 8

8. Sketch two triangles that have a correspondence. Describe the correspondence in symbols or words. Have a partner check your work.

Lesson Summary

- Two triangles and their respective parts can be compared once a correspondence has been assigned to the two triangles. Once a correspondence is selected, corresponding sides and corresponding angles can also be determined.

- Double arrows notate corresponding vertices. Triangle correspondences can also be notated with double arrows.

- Triangles are identical if there is a correspondence so that corresponding sides and angles are equal.

- An equal number of tick marks on two different sides indicates the sides are equal in measurement. An equal number of arcs on two different angles indicates the angles are equal in measurement.

Problem Set

Given the following triangle correspondences, use double arrows to show the correspondence between vertices, angles, and sides.

1.

Triangle Correspondence	△ ABC ↔ △ RTS
Correspondence of Vertices	
Correspondence of Angles	
Correspondence of Sides	

2.

Triangle Correspondence	△ ABC ↔ △ FGE
Correspondence of Vertices	
Correspondence of Angles	
Correspondence of Sides	

EUREKA
MATH™

Lesson 5:   Identical Triangles

3.

Triangle Correspondence	$\triangle QRP \leftrightarrow \triangle WYX$
Correspondence of Vertices	
Correspondence of Angles	
Correspondence of Sides	

Name the angle pairs and side pairs to find a triangle correspondence that matches sides of equal length and angles of equal measurement.

4.

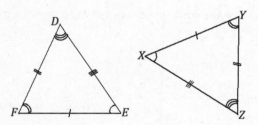

5.

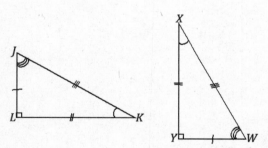

6.

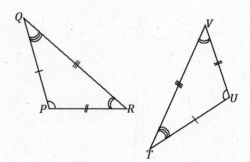

EUREKA
MATH™

7. Consider the following points in the coordinate plane.

    a. How many different (non-identical) triangles can be drawn using any three of these six points as vertices?

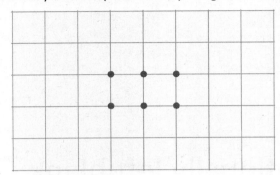

    b. How can we be sure that there are no more possible triangles?

8. Quadrilateral $ABCD$ is identical with quadrilateral $WXYZ$ with a correspondence $A \leftrightarrow W, B \leftrightarrow X, C \leftrightarrow Y$, and $D \leftrightarrow Z$.

    a. In the figure below, label points W, X, Y, and Z on the second quadrilateral.

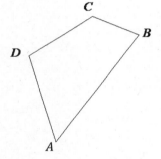 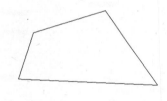

    b. Set up a correspondence between the side lengths of the two quadrilaterals that matches sides of equal length.

    c. Set up a correspondence between the angles of the two quadrilaterals that matches angles of equal measure.

This page intentionally left blank

Lesson 6: Drawing Geometric Shapes

Classwork

Exploratory Challenge

Use a ruler, protractor, and compass to complete the following problems.

1. Use your ruler to draw three segments of the following lengths: 4 cm, 7.2 cm, and 12.8 cm. Label each segment with its measurement.

2. Draw complementary angles so that one angle is 35°. Label each angle with its measurement. Are the angles required to be adjacent?

3. Draw vertical angles so that one angle is 125°. Label each angle formed with its measurement.

©2015 Great Minds. eureka-math.org
G7-M6-SE-B3-1.3.1-01.2016

4. Draw three distinct segments of lengths 2 cm, 4 cm, and 6 cm. Use your compass to draw three circles, each with a radius of one of the drawn segments. Label each radius with its measurement.

5. Draw three adjacent angles a, b, and c so that $a = 25°$, $b = 90°$, and $c = 50°$. Label each angle with its measurement.

6. Draw a rectangle $ABCD$ so that $AB = CD = 8$ cm and $BC = AD = 3$ cm.

©2015 Great Minds. eureka-math.org
G7-M6-SE-B3-1.3.1-01.2016

7. Draw a segment AB that is 5 cm in length. Draw a second segment that is longer than $\overline{AB}$, and label one endpoint C. Use your compass to find a point on your second segment, which will be labeled D, so that $CD = AB$.

8. Draw a segment AB with a length of your choice. Use your compass to construct two circles:

    i.   A circle with center A and radius AB.

    ii. A circle with center B and radius BA.

    Describe the construction in a sentence.

9. Draw a horizontal segment AB, 12 cm in length.

    a. Label a point O on $\overline{AB}$ that is 4 cm from B.

    b. Point O will be the vertex of an angle COB.

    c. Draw ray OC so that the ray is above $\overline{AB}$ and $\angle COB = 30°$.

    d. Draw a point P on $\overline{AB}$ that is 4 cm from A.

    e. Point P will be the vertex of an angle QPO.

    f. Draw ray PQ so that the ray is above $\overline{AB}$ and $\angle QPO = 30°$.

10. Draw segment AB of length 4 cm. Draw two circles that are the same size, one with center A and one with center B (i.e., do not adjust your compass in between) with a radius of a length that allows the two circles to intersect in two distinct locations. Label the points where the two circles intersect C and D. Join A and C with a segment; join B and C with a segment. Join A and D with a segment; join B and D with a segment.

   What kind of triangles are $\triangle ABC$ and $\triangle ABD$? Justify your response.

11. Determine all possible measurements in the following triangle, and use your tools to create a copy of it.

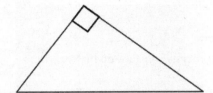

Lesson Summary

The compass is a tool that can be used for many purposes that include the following:

- Constructing circles.
- Measuring and marking a segment of equal length to another segment.
- Confirming that the radius of the center of a circle to the circle itself remains constant no matter where you are on the circle.

Problem Set

Use a ruler, protractor, and compass to complete the following problems.

1. Draw a segment AB that is 5 cm in length and perpendicular to segment CD, which is 2 cm in length.

2. Draw supplementary angles so that one angle is 26°. Label each angle with its measurement.

3. Draw △ ABC so that ∠B has a measurement of 100°.

4. Draw a segment AB that is 3 cm in length. Draw a circle with center A and radius AB. Draw a second circle with diameter AB.

5. Draw an isosceles △ ABC. Begin by drawing ∠A with a measurement of 80°. Use the rays of ∠A as the equal legs of the triangle. Choose a length of your choice for the legs, and use your compass to mark off each leg. Label each marked point with B and C. Label all angle measurements.

6. Draw an isosceles △ DEF. Begin by drawing a horizontal segment DE that is 6 cm in length. Use your protractor to draw ∠D and ∠E so that the measurements of both angles are 30°. If the non-horizontal rays of ∠D and ∠E do not already cross, extend each ray until the two rays intersect. Label the point of intersection F. Label all side and angle measurements.

7. Draw a segment AB that is 7 cm in length. Draw a circle with center A and a circle with center B so that the circles are not the same size, but do intersect in two distinct locations. Label one of these intersections C. Join A to C and B to C to form △ ABC.

8. Draw an isosceles trapezoid $WXYZ$ with two equal base angles, ∠W and ∠X, that each measures 110°. Use your compass to create the two equal sides of the trapezoid. Leave arc marks as evidence of the use of your compass. Label all angle measurements. Explain how you constructed the trapezoid.

Lesson 7: Drawing Parallelograms

Classwork

Example 1

Use what you know about drawing parallel lines with a setsquare to draw rectangle $ABCD$ with dimensions of your choice. State the steps you used to draw your rectangle, and compare those steps to those of a partner.

Example 2

Use what you know about drawing parallel lines with a setsquare to draw rectangle $ABCD$ with $AB = 3$ cm and $BC = 5$ cm. Write a plan for the steps you will take to draw $ABCD$.

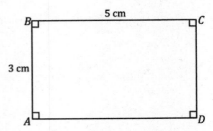

Example 3

Use a setsquare, ruler, and protractor to draw parallelogram $PQRS$ so that the measurement of $\angle P$ is 50°, $PQ = 5$ cm, the measurement of $\angle Q$ is 130°, and the length of the altitude to $\overline{PQ}$ is 4 cm.

Exercise 1

Use a setsquare, ruler, and protractor to draw parallelogram $DEFG$ so that the measurement of $\angle D$ is 40°, $DE = 3$ cm, the measurement of $\angle E$ is 140°, and the length of the altitude to $\overline{DE}$ is 5 cm.

Example 4

Use a setsquare, ruler, and protractor to draw rhombus $ABCD$ so that the measurement of $\angle A = 80°$, the measurement of $\angle B = 100°$, and each side of the rhombus measures 5 cm.

Lesson Summary

A protractor, ruler, and setsquare are necessary tools to construct a parallelogram. A setsquare is the tool that gives a means to draw parallel lines for the sides of a parallelogram.

Problem Set

1. Draw rectangle $ABCD$ with $AB = 5$ cm and $BC = 7$ cm.

2. Use a setsquare, ruler, and protractor to draw parallelogram $PQRS$ so that the measurement of $\angle P$ is $65°$, $PQ = 8$ cm, the measurement of $\angle Q$ is $115°$, and the length of the altitude to $\overline{PQ}$ is 3 cm.

3. Use a setsquare, ruler, and protractor to draw rhombus $ABCD$ so that the measurement of $\angle A$ is $60°$, and each side of the rhombus measures 5 cm.

The following table contains partial information for parallelogram $ABCD$. Using no tools, make a sketch of the parallelogram. Then, use a ruler, protractor, and setsquare to draw an accurate picture. Finally, complete the table with the unknown lengths.

	$\angle A$	AB	Altitude to $\overline{AB}$	BC	Altitude to $\overline{BC}$
4.	$45°$	5 cm		4 cm	
5.	$50°$	3 cm		3 cm	
6.	$60°$	4 cm	4 cm		

7. Use what you know about drawing parallel lines with a setsquare to draw trapezoid $ABCD$ with parallel sides $\overline{AB}$ and $\overline{CD}$. The length of $\overline{AB}$ is 3 cm, and the length of $\overline{CD}$ is 5 cm; the height between the parallel sides is 4 cm. Write a plan for the steps you will take to draw $ABCD$.

8. Use the appropriate tools to draw rectangle $FIND$ with $FI = 5$ cm and $IN = 10$ cm.

9. Challenge: Determine the area of the largest rectangle that will fit inside an equilateral triangle with side length 5 cm.

Lesson 8: Drawing Triangles

Classwork

Exercises 1–2

1. Use your protractor and ruler to draw right triangle DEF. Label all sides and angle measurements.

    a. Predict how many of the right triangles drawn in class are identical to the triangle you have drawn.

    b. How many of the right triangles drawn in class are identical to the triangle you drew? Were you correct in your prediction?

2. Given the following three sides of $\triangle ABC$, use your compass to copy the triangle. The longest side has been copied for you already. Label the new triangle $A'B'C'$, and indicate all side and angle measurements. For a reminder of how to begin, refer to Lesson 6 Exploratory Challenge Problem 10.

    A_____B

    B_____C

    A_____C

    A_____C

Exploratory Challenge

A triangle is to be drawn provided the following conditions: the measurements of two angles are 30° and 60°, and the length of a side is 10 cm. Note that where each of these measurements is positioned is not fixed.

    a.   How is the premise of this problem different from Exercise 2?

    b.   Given these measurements, do you think it will be possible to draw more than one triangle so that the triangles drawn will be different from each other? Or do you think attempting to draw more than one triangle with these measurements will keep producing the same triangle, just turned around or flipped about?

    c.   Based on the provided measurements, draw $\triangle ABC$ so that $\angle A = 30°$, $\angle B = 60°$, and $AB = 10$ cm. Describe how the 10 cm side is positioned.

d. Now, using the same measurements, draw △ $A'B'C'$ so that $\angle A' = 30°$, $\angle B' = 60°$, and $AC = 10$ cm. Describe how the 10 cm side is positioned.

e. Lastly, again, using the same measurements, draw △ $A''B''C''$ so that $\angle A'' = 30°$, $\angle B'' = 60°$, and $B''C'' = 10$ cm. Describe how the 10 cm side is positioned.

f. Are the three drawn triangles identical? Justify your response using measurements.

g.   Draw $\triangle\, A'''B'''C'''$ so that $\angle B''' = 30°$, $\angle C''' = 60°$, and $B'''C''' = 10$ cm. Is it identical to any of the three triangles already drawn?

h.   Draw another triangle that meets the criteria of this challenge. Is it possible to draw any other triangles that would be different from the three drawn above?

©2015 Great Minds. eureka-math.org
G7-M6-SE-B3-1.3.1-01.2016

Lesson Summary

The following conditions produce identical triangles:

What Criteria Produce Unique Triangles?

Criteria	Example

Problem Set

1. Draw three different acute triangles XYZ, $X'Y'Z'$, and $X''Y''Z''$ so that one angle in each triangle is 45°. Label all sides and angle measurements. Why are your triangles not identical?

2. Draw three different equilateral triangles ABC, $A'B'C'$, and $A''B''C''$. A side length of $\triangle ABC$ is 3 cm. A side length of $\triangle A'B'C'$ is 5 cm. A side length of $\triangle A''B''C''$ is 7 cm. Label all sides and angle measurements. Why are your triangles not identical?

3. Draw as many isosceles triangles that satisfy the following conditions: one angle measures 110°, and one side measures 6 cm. Label all angle and side measurements. How many triangles can be drawn under these conditions?

4. Draw three nonidentical triangles so that two angles measure 50° and 60° and one side measures 5 cm.

    a. Why are the triangles not identical?

    b. Based on the diagrams you drew for part (a) and for Problem 2, what can you generalize about the criterion of three given angles in a triangle? Does this criterion determine a unique triangle?

This page intentionally left blank

Lesson 9: Conditions for a Unique Triangle—Three Sides and Two Sides and the Included Angle

Classwork

Exploratory Challenge

1. A triangle XYZ exists with side lengths of the segments below. Draw $\triangle X'Y'Z'$ with the same side lengths as $\triangle XYZ$. Use your compass to determine the sides of $\triangle X'Y'Z'$. Use your ruler to measure side lengths. Leave all construction marks as evidence of your work, and label all side and angle measurements.

    Under what condition is $\triangle X'Y'Z'$ drawn? Compare the triangle you drew to two of your peers' triangles. Are the triangles identical? Did the condition determine a unique triangle? Use your construction to explain why. Do the results differ from your predictions?

    X _____ Y

    Y _____ Z

    X _____ Z

2. $\triangle ABC$ is located below. Copy the sides of the triangle to create $\triangle A'B'C'$. Use your compass to determine the sides of $\triangle A'B'C'$. Use your ruler to measure side lengths. Leave all construction marks as evidence of your work, and label all side and angle measurements.

   Under what condition is $\triangle A'B'C'$ drawn? Compare the triangle you drew to two of your peers' triangles. Are the triangles identical? Did the condition determine a unique triangle? Use your construction to explain why.

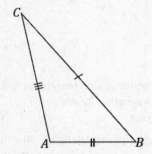

EUREKA MATH™

3. A triangle DEF has an angle of 40° adjacent to side lengths of 4 cm and 7 cm. Construct △ $D'E'F'$ with side lengths $D'E' = 4$ cm, $D'F' = 7$ cm, and included angle $\angle D' = 40°$. Use your compass to draw the sides of △ $D'E'F'$. Use your ruler to measure side lengths. Leave all construction marks as evidence of your work, and label all side and angle measurements.

Under what condition is △ $D'E'F'$ drawn? Compare the triangle you drew to two of your peers' triangles. Did the condition determine a unique triangle? Use your construction to explain why.

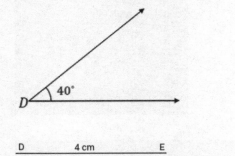

EUREKA
MATH™

Lesson 9:   Conditions for a Unique Triangle—Three Sides and Two Sides and the
                Included Angle

S.61

©2015 Great Minds. eureka-math.org
G7-M6-SE-B3-1.3.1-01.2016

4. $\triangle XYZ$ has side lengths $XY = 2.5$ cm, $XZ = 4$ cm, and $\angle X = 120°$. Draw $\triangle X'Y'Z'$ under the same conditions. Use your compass and protractor to draw the sides of $\triangle X'Y'Z'$. Use your ruler to measure side lengths. Leave all construction marks as evidence of your work, and label all side and angle measurements.

Under what condition is $\triangle X'Y'Z'$ drawn? Compare the triangle you drew to two of your peers' triangles. Are the triangles identical? Did the condition determine a unique triangle? Use your construction to explain why.

©2015 Great Minds. eureka-math.org
G7-M6-SE-B3-1.3.1-01.2016

Lesson Summary

The following conditions determine a unique triangle:

- Three sides.
- Two sides and an included angle.

Problem Set

1. A triangle with side lengths 3 cm, 4 cm, and 5 cm exists. Use your compass and ruler to draw a triangle with the same side lengths. Leave all construction marks as evidence of your work, and label all side and angle measurements.

   Under what condition is the triangle drawn? Compare the triangle you drew to two of your peers' triangles. Are the triangles identical? Did the condition determine a unique triangle? Use your construction to explain why.

2. Draw triangles under the conditions described below.

   a. A triangle has side lengths 5 cm and 6 cm. Draw two nonidentical triangles that satisfy these conditions. Explain why your triangles are not identical.

   b. A triangle has a side length of 7 cm opposite a 45° angle. Draw two nonidentical triangles that satisfy these conditions. Explain why your triangles are not identical.

3. Diagonal $\overline{BD}$ is drawn in square $ABCD$. Describe what condition(s) can be used to justify that $\triangle ABD$ is identical to $\triangle CBD$. What can you say about the measures of $\angle ABD$ and $\angle CBD$? Support your answers with a diagram and explanation of the correspondence(s) that exists.

4. Diagonals $\overline{BD}$ and $\overline{AC}$ are drawn in square $ABCD$. Show that $\triangle ABC$ is identical to $\triangle BAD$, and then use this information to show that the diagonals are equal in length.

5. Diagonal $\overline{QS}$ is drawn in rhombus $PQRS$. Describe the condition(s) that can be used to justify that $\triangle PQS$ is identical to $\triangle RQS$. Can you conclude that the measures of $\angle PQS$ and $\angle RQS$ are the same? Support your answer with a diagram and explanation of the correspondence(s) that exists.

6. Diagonals $\overline{QS}$ and $\overline{PR}$ are drawn in rhombus $PQRS$ and meet at point T. Describe the condition(s) that can be used to justify that $\triangle PQT$ is identical to $\triangle RQT$. Can you conclude that the line segments PR and QS are perpendicular to each other? Support your answers with a diagram and explanation of the correspondence(s) that exists.

This page intentionally left blank

Lesson 10: Conditions for a Unique Triangle—Two Angles and a Given Side

Classwork

Exploratory Challenge

1. A triangle XYZ has angle measures $\angle X = 30°$ and $\angle Y = 50°$ and included side $XY = 6$ cm. Draw $\triangle X'Y'Z'$ under the same condition as $\triangle XYZ$. Leave all construction marks as evidence of your work, and label all side and angle measurements.

    Under what condition is $\triangle X'Y'Z'$ drawn? Compare the triangle you drew to two of your peers' triangles. Are the triangles identical? Did the condition determine a unique triangle? Use your construction to explain why.

2. A triangle RST has angle measures $\angle S = 90°$ and $\angle T = 45°$ and included side $ST = 7$ cm. Draw $\triangle R'S'T'$ under the same condition. Leave all construction marks as evidence of your work, and label all side and angle measurements.

    Under what condition is $\triangle R'S'T'$ drawn? Compare the triangle you drew to two of your peers' triangles. Are the triangles identical? Did the condition determine a unique triangle? Use your construction to explain why.

3. A triangle JKL has angle measures $\angle J = 60°$ and $\angle L = 25°$ and side $KL = 5$ cm. Draw $\triangle J'K'L'$ under the same condition. Leave all construction marks as evidence of your work, and label all side and angle measurements.

    Under what condition is $\triangle J'K'L'$ drawn? Compare the triangle you drew to two of your peers' triangles. Are the triangles identical? Did the condition determine a unique triangle? Use your construction to explain why.

Lesson 10: Conditions for a Unique Triangle—Two Angles and a Given Side

EUREKA MATH

4. A triangle ABC has angle measures $\angle C = 35°$ and $\angle B = 105°$ and side $AC = 7$ cm. Draw $\triangle\, A'B'C'$ under the same condition. Leave all construction marks as evidence of your work, and label all side and angle measurements.

Under what condition is $\triangle\, A'B'C'$ drawn? Compare the triangle you drew to two of your peers' triangles. Are the triangles identical? Did the condition determine a unique triangle? Use your construction to explain why.

Lesson Summary

The following conditions determine a unique triangle:

- Three sides.
- Two sides and included angle.
- Two angles and the included side.
- Two angles and the side opposite.

Problem Set

1. In $\triangle FGH$, $\angle F = 42°$ and $\angle H = 70°$. $FH = 6$ cm. Draw $\triangle F'G'H'$ under the same condition as $\triangle FGH$. Leave all construction marks as evidence of your work, and label all side and angle measurements.

   What can you conclude about $\triangle FGH$ and $\triangle F'G'H'$? Justify your response.

2. In $\triangle WXY$, $\angle Y = 57°$ and $\angle W = 103°$. Side $YX = 6.5$ cm. Draw $\triangle W'X'Y'$ under the same condition as $\triangle WXY$. Leave all construction marks as evidence of your work, and label all side and angle measurements.

   What can you conclude about $\triangle WXY$ and $\triangle W'X'Y'$? Justify your response.

3. Points A, Z, and E are collinear, and $\angle B = \angle D$. What can be concluded about $\triangle ABZ$ and $\triangle EDZ$? Justify your answer.

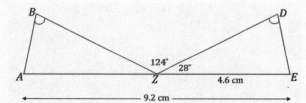

4. Draw $\triangle ABC$ so that $\angle A$ has a measurement of $60°$, $\angle B$ has a measurement of $60°$, and $\overline{AB}$ has a length of 8 cm. What are the lengths of the other sides?

5. Draw $\triangle ABC$ so that $\angle A$ has a measurement of $30°$, $\angle B$ has a measurement of $60°$, and $\overline{BC}$ has a length of 5 cm. What is the length of the longest side?

EUREKA
MATH™

Lesson 11: Conditions on Measurements That Determine a Triangle

Classwork

Exploratory Challenge 1

a.   Can any three side lengths form a triangle? Why or why not?

b.   Draw a triangle according to these instructions:
   ✓   Draw segment AB of length 10 cm in your notebook.
   ✓   Draw segment BC of length 5 cm on one piece of patty paper.
   ✓   Draw segment AC of length 3 cm on the other piece of patty paper.
   ✓   Line up the appropriate endpoint on each piece of patty paper with the matching endpoint on segment AB.
   ✓   Use your pencil point to hold each patty paper in place, and adjust the paper to form $\triangle ABC$.

c.   What do you notice?

d.   What must be true about the sum of the lengths of $\overline{AC}$ and $\overline{BC}$ if the two segments were to just meet? Use your patty paper to verify your answer.

e.   Based on your conclusion for part (d), what if $AC = 3$ cm as you originally had, but $BC = 10$ cm. Could you form $\triangle ABC$?

f.   What must be true about the sum of the lengths of $\overline{AC}$ and $\overline{BC}$ if the two segments were to meet and form a triangle?

Exercise 1

Two sides of $\triangle DEF$ have lengths of 5 cm and 8 cm. What are all the possible whole number lengths for the remaining side?

EUREKA
MATH™

Exploratory Challenge 2

a. Which of the following conditions determine a triangle? Follow the instructions to try to draw △ ABC.
    Segment AB has been drawn for you as a starting point in each case.

    i. Choose measurements of ∠A and ∠B for △ ABC so that the sum of measurements is greater than 180°.
        Label your diagram.

        Your chosen angle measurements: ∠A =   ∠B =

        Were you able to form a triangle? Why or why not?

        A ————————————————————————————— B

    ii. Choose measurements of ∠A and ∠B for △ ABC so that the measurement of ∠A is supplementary to the
        measurement of ∠B. Label your diagram.

        Your chosen angle measurements: ∠A =   ∠B =

        Were you able to form a triangle? Why or why not?

        A ————————————————————————————— B

©2015 Great Minds. eureka-math.org
G7-M6-SE-B3-1.3.1-01.2016

iii. Choose measurements of ∠A and ∠B for △ ABC so that the sum of measurements is less than 180°. Label your diagram.

Your chosen angle measurements: ∠A =   ∠B =

Were you able to form a triangle? Why or why not?

A ——————————————————— B

b. Which condition must be true regarding angle measurements in order to determine a triangle?

c. Measure and label the formed triangle in part (a) with all three side lengths and the angle measurement for ∠C. Now, use a protractor, ruler, and compass to draw △ A'B'C' with the same angle measurements but side lengths that are half as long.

d.   Do the three angle measurements of a triangle determine a unique triangle? Why or why not?

Exercise 2

Which of the following sets of angle measurements determines a triangle?

a.   30°, 120°

b.   125°, 55°

c.   105°, 80°

d.   90°, 89°

e.   91°, 89°

Choose one example from above that does determine a triangle and one that does not. For each, explain why it does or does not determine a triangle using words and a diagram.

Lesson Summary

- Three lengths determine a triangle provided the largest length is less than the sum of the other two lengths.
- Two angle measurements determine a triangle provided the sum of the two angle measurements is less than 180°.
- Three given angle measurements do not determine a unique triangle.
- Scale drawings of a triangle have equal corresponding angle measurements, but corresponding side lengths are proportional.

Problem Set

1. Decide whether each set of three given lengths determines a triangle. For any set of lengths that does determine a triangle, use a ruler and compass to draw the triangle. Label all side lengths. For sets of lengths that do not determine a triangle, write "Does not determine a triangle," and justify your response.

    a. 3 cm, 4 cm, 5 cm

    b. 1 cm, 4 cm, 5 cm

    c. 1 cm, 5 cm, 5 cm

    d. 8 cm, 3 cm, 4 cm

    e. 8 cm, 8 cm, 4 cm

    f. 4 cm, 4 cm, 4 cm

2. For each angle measurement below, provide one angle measurement that will determine a triangle and one that will not determine a triangle. Provide a brief justification for the angle measurements that will not form a triangle. Assume that the angles are being drawn to a horizontal segment AB; describe the position of the non-horizontal rays of angles $\angle A$ and $\angle B$.

$\angle A$	$\angle B$: A Measurement That Determines a Triangle	$\angle B$: A Measurement That *Does Not* Determine a Triangle	Justification for No Triangle
40°			
100°			
90°			
135°			

©2015 Great Minds. eureka-math.org
G7-M6-SE-B3-1.3.1-01.2016

3. For the given side lengths, provide the minimum and maximum whole number side lengths that determine a triangle.

Given Side Lengths	Minimum Whole Number Third Side Length	Maximum Whole Number Third Side Length
5 cm, 6 cm		
3 cm, 7 cm		
4 cm, 10 cm		
1 cm, 12 cm		

This page intentionally left blank

Lesson 12: Unique Triangles—Two Sides and a Non-Included Angle

Classwork

Exploratory Challenge

1. Use your tools to draw $\triangle ABC$ in the space below, provided $AB = 5$ cm, $BC = 3$ cm, and $\angle A = 30°$. Continue with the rest of the problem as you work on your drawing.

a.   What is the relationship between the given parts of △ ABC?

b.   Which parts of the triangle can be drawn without difficulty? What makes this drawing challenging?

c.   A ruler and compass are instrumental in determining where C is located.
   ✓   Even though the length of segment AC is unknown, extend the ray AC in anticipation of the intersection with segment BC.
   ✓   Draw segment BC with length 3 cm away from the drawing of the triangle.
   ✓   Adjust your compass to the length of $\overline{BC}$.
   ✓   Draw a circle with center B and a radius equal to BC, or 3 cm.

d.   How many intersections does the circle make with segment AC? What does each intersection signify?

e.   Complete the drawing of △ ABC.

f.   Did the results of your drawing differ from your prediction?

2. Now attempt to draw $\triangle DEF$ in the space below, provided $DE = 5$ cm, $EF = 3$ cm, and $\angle F = 90°$. Continue with the rest of the problem as you work on your drawing.

a. How are these conditions different from those in Exercise 1, and do you think the criteria will determine a unique triangle?

b. What is the relationship between the given parts of $\triangle DEF$?

c. Describe how you will determine the position of $\overline{DE}$.

d. How many intersections does the circle make with $\overline{FD}$?

e. Complete the drawing of $\triangle DEF$. How is the outcome of $\triangle DEF$ different from that of $\triangle ABC$?

f. Did your results differ from your prediction?

Lesson 12:   Unique Triangles—Two Sides and a Non-Included Angle

EUREKA MATH™

3. Now attempt to draw △ JKL, provided $KL = 8$ cm, $KJ = 4$ cm, and $\angle J = 120°$. Use what you drew in Exercises 1 and 2 to complete the full drawing.

4. Review the conditions provided for each of the three triangles in the Exploratory Challenge, and discuss the uniqueness of the resulting drawing in each case.

Problem Set

1. In each of the triangles below, two sides and a non-included acute angle are marked. Use a compass to draw a nonidentical triangle that has the same measurements as the marked angle and marked sides (look at Exercise 1, part (e) of the Exploratory Challenge as a reference). Draw the new triangle on top of the old triangle. What is true about the marked angles in each triangle that results in two non-identical triangles under this condition?

    a.

    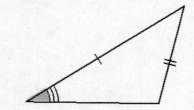

    b.

    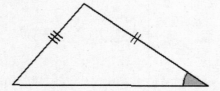

    c.

    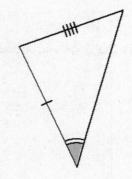

EUREKA
MATH™

2. Sometimes two sides and a non-included angle of a triangle determine a unique triangle, even if the angle is acute. In the following two triangles, copy the marked information (i.e., two sides and a non-included acute angle), and discover which determines a unique triangle. Measure and label the marked parts.

In each triangle, how does the length of the marked side adjacent to the marked angle compare with the length of the side opposite the marked angle? Based on your drawings, specifically state when the two sides and acute non-included angle condition determines a unique triangle.

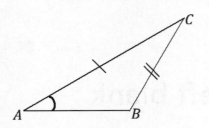   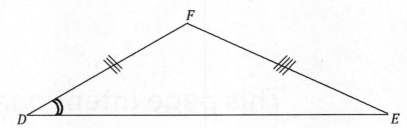

3. A sub-condition of the two sides and non-included angle is provided in each row of the following table. Decide whether the information determines a unique triangle. Answer with a *yes*, *no*, or *maybe* (for a case that may or may not determine a unique triangle).

	Condition	Determines a Unique Triangle?
1	Two sides and a non-included 90° angle.	
2	Two sides and an acute, non-included angle.	
3	Two sides and a non-included 140° angle.	
4	Two sides and a non-included 20° angle, where the side adjacent to the angle is shorter than the side opposite the angle.	
5	Two sides and a non-included angle.	
6	Two sides and a non-included 70° angle, where the side adjacent to the angle is longer than the side opposite the angle.	

4. Choose one condition from the table in Problem 3 that does not determine a unique triangle, and explain why.

5. Choose one condition from the table in Problem 3 that does determine a unique triangle, and explain why.

This page intentionally left blank

Lesson 13: Checking for Identical Triangles

Classwork

Opening Exercise

    a.   List all the conditions that determine unique triangles.

    b.   How are the terms *identical* and *unique* related?

Each of the following problems gives two triangles. State whether the triangles are *identical*, *not identical*, or *not necessarily identical*. If the triangles are identical, give the triangle conditions that explain why, and write a triangle correspondence that matches the sides and angles. If the triangles are not identical, explain why. If it is not possible to definitively determine whether the triangles are identical, write "the triangles are not necessarily identical," and explain your reasoning.

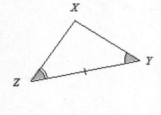

> Example 1

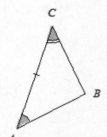

Exercises 1–3

1.

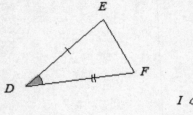

2.

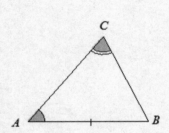 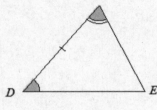

3.

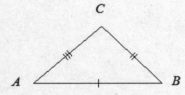 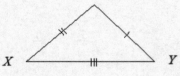

EUREKA MATH

In Example 2 and Exercises 4–6, three pieces of information are given for $\triangle ABC$ and $\triangle XYZ$. Draw, freehand, the two triangles (do not worry about scale), and mark the given information. If the triangles are identical, give a triangle correspondence that matches equal angles and equal sides. Explain your reasoning.

Example 2

$AB = XZ, AC = XY, \angle A = \angle X$

Exercises 4–6

4. $\angle A = \angle Z, \angle B = \angle Y, AB = YZ$

5. $\angle A = \angle Z, \angle B = \angle Y, BC = XY$

6. $\angle A = \angle Z, \angle B = \angle Y, BC = XZ$

Lesson Summary

The measurement and arrangement (and correspondence) of the parts in each triangle play a role in determining whether two triangles are identical.

Problem Set

In each of the following four problems, two triangles are given. State whether the triangles are *identical*, *not identical*, or *not necessarily identical*. If the triangles are identical, give the triangle conditions that explain why, and write a triangle correspondence that matches the sides and angles. If the triangles are not identical, explain why. If it is not possible to definitively determine whether the triangles are identical, write "the triangles are not necessarily identical," and explain your reasoning.

1.

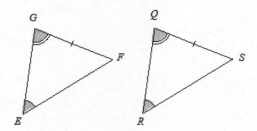

2.

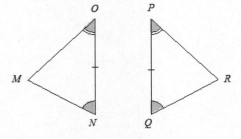

3.

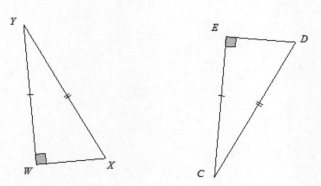

4.

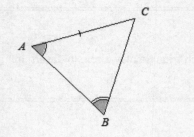 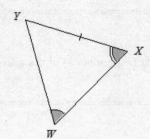

For Problems 5–8, three pieces of information are given for △ *ABC* and △ *YZX*. Draw, freehand, the two triangles (do not worry about scale), and mark the given information. If the triangles are identical, give a triangle correspondence that matches equal angles and equal sides. Explain your reasoning.

5. $AB = YZ, BC = ZX, AC = YX$

6. $AB = YZ, BC = ZX, \angle C = \angle Y$

7. $AB = XZ, \angle A = \angle Z, \angle C = \angle Y$

8. $AB = XY, AC = YZ, \angle C = \angle Z$ (Note that both angles are obtuse.)

EUREKA
MATH™

Lesson 14: Checking for Identical Triangles

Classwork

In each of the following problems, determine whether the triangles are *identical*, *not identical*, or *not necessarily identical*; justify your reasoning. If the relationship between the two triangles yields information that establishes a condition, describe the information. If the triangles are identical, write a triangle correspondence that matches the sides and angles.

Example 1

What is the relationship between the two triangles below?

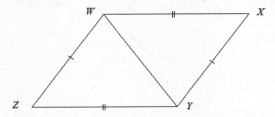

Exercises 1–2

1. Are the triangles identical? Justify your reasoning.

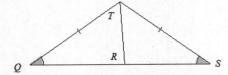

2. Are the triangles identical? Justify your reasoning.

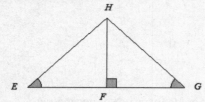

Example 2

Are the triangles identical? Justify your reasoning.

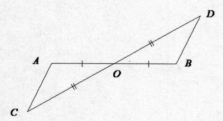

Lesson 14: Checking for Identical Triangles

Exercises 3–4

3.   Are the triangles identical? Justify your reasoning.

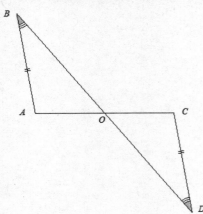

4.   Are the triangles identical? Justify your reasoning.

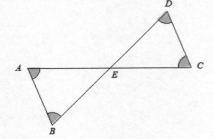

Exercises 5–8

5. Are the triangles identical? Justify your reasoning.

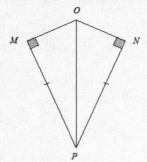

6. Are the triangles identical? Justify your reasoning.

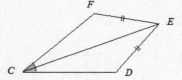

Lesson 14:   Checking for Identical Triangles

7. Are the triangles identical? Justify your reasoning.

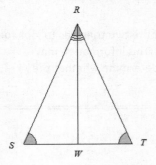

8. Create your own labeled diagram and set of criteria for a pair of triangles. Ask a neighbor to determine whether the triangles are identical based on the provided information.

Lesson Summary

In deciding whether two triangles are identical, examine the structure of the diagram of the two triangles to look for a relationship that might reveal information about corresponding parts of the triangles. This information may determine whether the parts of the triangle satisfy a particular condition, which might determine whether the triangles are identical.

Problem Set

In the following problems, determine whether the triangles are *identical*, *not identical*, or *not necessarily identical*; justify your reasoning. If the relationship between the two triangles yields information that establishes a condition, describe the information. If the triangles are identical, write a triangle correspondence that matches the sides and angles.

1.
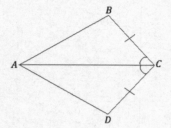

2.
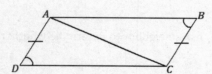

3.
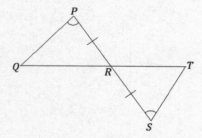

©2015 Great Minds. eureka-math.org
G7-M6-SE-B3-1.3.1-01.2016

4.

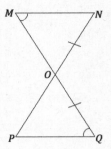

5.

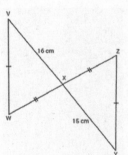

6.

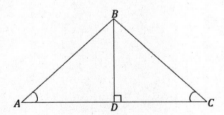

7.

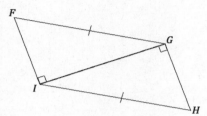

8. Are there any identical triangles in this diagram?

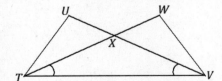

9.

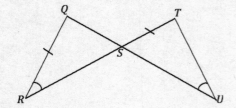

10.

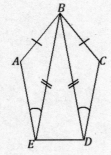

EUREKA
MATH™

Lesson 15: Using Unique Triangles to Solve Real-World and Mathematical Problems

Classwork

Example 1

A triangular fence with two equal angles, $\angle S = \angle T$, is used to enclose some sheep. A fence is constructed inside the triangle that exactly cuts the other angle into two equal angles: $\angle SRW = \angle TRW$. Show that the gates, represented by $\overline{SW}$ and $\overline{WT}$, are the same width.

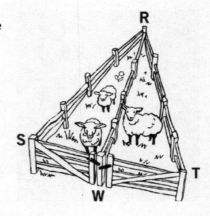

Example 2

In $\triangle ABC$, $AC = BC$, and $\triangle ABC \leftrightarrow \triangle B'A'C'$. John says that the triangle correspondence matches two sides and the included angle and shows that $\angle A = \angle B'$. Is John correct?

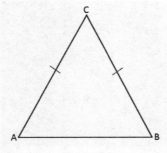
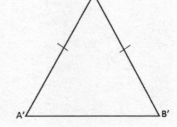

Exercises 1–4

1. Mary puts the center of her compass at the vertex O of the angle and locates points A and B on the sides of the angle. Next, she centers her compass at each of A and B to locate point C. Finally, she constructs the ray $\overrightarrow{OC}$. Explain why $\angle BOC = \angle AOC$.

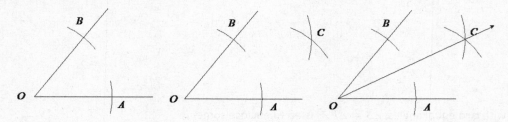

2. Quadrilateral $ACBD$ is a model of a kite. The diagonals $\overline{AB}$ and $\overline{CD}$ represent the sticks that help keep the kite rigid.

   a. John says that $\angle ACD = \angle BCD$. Can you use identical triangles to show that John is correct?

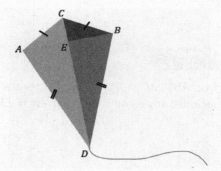

b.   Jill says that the two sticks are perpendicular to each other. Use the fact that $\angle ACD = \angle BCD$ and what you know about identical triangles to show $\angle AEC = 90°$.

c.   John says that Jill's triangle correspondence that shows the sticks are perpendicular to each other also shows that the sticks cross at the midpoint of the horizontal stick. Is John correct? Explain.

3.   In $\triangle ABC$, $\angle A = \angle B$, and $\triangle ABC \leftrightarrow \triangle B'A'C'$. Jill says that the triangle correspondence matches two angles and the included side and shows that $AC = B'C'$. Is Jill correct?

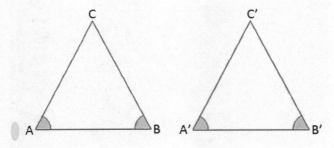

4. Right triangular corner flags are used to mark a soccer field. The vinyl flags have a base of 40 cm and a height of 14 cm.

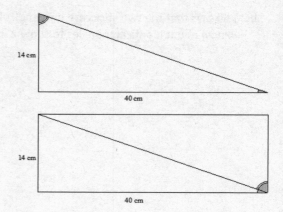

a. Mary says that the two flags can be obtained by cutting a rectangle that is 40 cm × 14 cm on the diagonal. Will that create two identical flags? Explain.

b. Will measures the two non-right angles on a flag and adds the measurements together. Can you explain, without measuring the angles, why his answer is 90°?

EUREKA MATH™

Lesson Summary

- In deciding whether two triangles are identical, examine the structure of the diagram of the two triangles to look for a relationship that might reveal information about corresponding parts of the triangles. This information may determine whether the parts of the triangle satisfy a particular condition, which might determine whether the triangles are identical.

- Be sure to identify and label all known measurements, and then determine if any other measurements can be established based on knowledge of geometric relationships.

Problem Set

1. Jack is asked to cut a cake into 8 equal pieces. He first cuts it into equal fourths in the shape of rectangles, and then he cuts each rectangle along a diagonal.

   Did he cut the cake into 8 equal pieces? Explain.

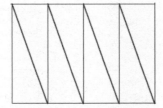

2. The bridge below, which crosses a river, is built out of two triangular supports. The point M lies on $\overline{BC}$. The beams represented by $\overline{AM}$ and $\overline{DM}$ are equal in length, and the beams represented by $\overline{AB}$ and $\overline{DC}$ are equal in length. If the supports were constructed so that $\angle A$ and $\angle D$ are equal in measurement, is point M the midpoint of $\overline{BC}$? Explain.

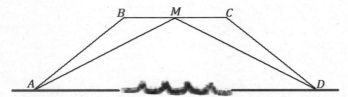

This page intentionally left blank

Lesson 16: Slicing a Right Rectangular Prism with a Plane

Classwork

Example 1

Consider a ball B. Figure 3 shows one possible slice of B.

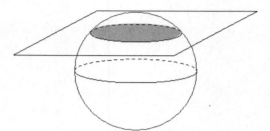

Figure 3. A Slice of Ball B

a. What figure does the slicing plane form? Students may choose
their method of representation of the slice (e.g., drawing a 2D
sketch, a 3D sketch, or describing the slice in words).

b. Will all slices that pass through B be the same size? Explain your reasoning.

c. How will the plane have to meet the ball so that the plane section consists of just one point?

Example 2

The right rectangular prism in Figure 4 has been sliced with a plane parallel to face $ABCD$. The resulting slice is a rectangular region that is identical to the parallel face.

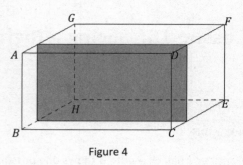

Figure 4

    a.   Label the vertices of the rectangular region defined by the slice as $WXYZ$.

    b.   To which other face is the slice parallel and identical?

    c.   Based on what you know about right rectangular prisms, which faces must the slice be perpendicular to?

Exercise 1

Discuss the following questions with your group.

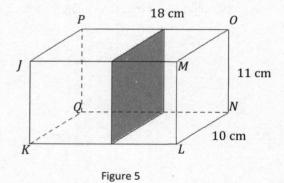

Figure 5

1.   The right rectangular prism in Figure 5 has been sliced with a plane parallel to face $LMON$.

    a.   Label the vertices of the rectangle defined by the slice as $RSTU$.

    b.   What are the dimensions of the slice?

    c.   Based on what you know about right rectangular prisms, which faces must the slice be perpendicular to?

EUREKA
MATH™

Example 3

The right rectangular prism in Figure 6 has been sliced with a plane perpendicular to $BCEH$. The resulting slice is a rectangular region with a height equal to the height of the prism.

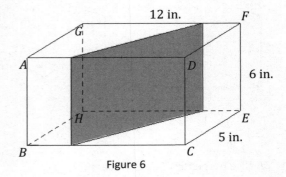

Figure 6

a.   Label the vertices of the rectangle defined by the slice as $WXYZ$.

b.   To which other face is the slice perpendicular?

c.   What is the height of rectangle $WXYZ$?

d.   Joey looks at $WXYZ$ and thinks that the slice may be a parallelogram that is not a rectangle. Based on what is known about how the slice is made, can he be right? Justify your reasoning.

Exercises 2–6

In the following exercises, the points at which a slicing plane meets the edges of the right rectangular prism have been marked. Each slice is either parallel or perpendicular to a face of the prism. Use a straightedge to join the points to outline the rectangular region defined by the slice, and shade in the rectangular slice.

2.   A slice parallel to a face

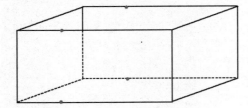

3. A slice perpendicular to a face

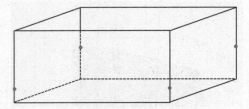

4. A slice perpendicular to a face

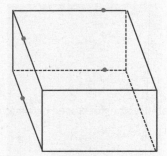

In Exercises 5–6, the dimensions of the prisms have been provided. Use the dimensions to sketch the slice from each prism, and provide the dimensions of each slice.

5. A slice parallel to a face

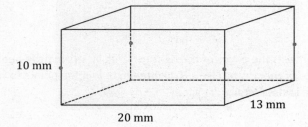

10 mm

20 mm   13 mm

6. A slice perpendicular to a face

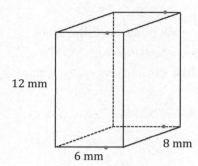

12 mm

6 mm

8 mm

Lesson Summary

- A slice, also known as a plane section, consists of all the points where the plane meets the figure.
- A slice made parallel to a face in a right rectangular prism will be parallel and identical to the face.
- A slice made perpendicular to a face in a right rectangular prism will be a rectangular region with a height equal to the height of the prism.

Problem Set

A right rectangular prism is shown along with line segments that lie in a face. For each line segment, draw and give the approximate dimensions of the slice that results when the slicing plane contains the given line segment and is perpendicular to the face that contains the line segment.

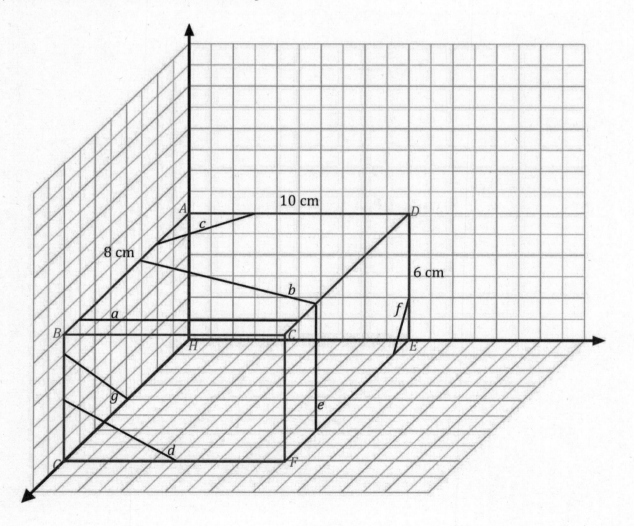

Lesson 16: Slicing a Right Rectangular Prism with a Plane

a.

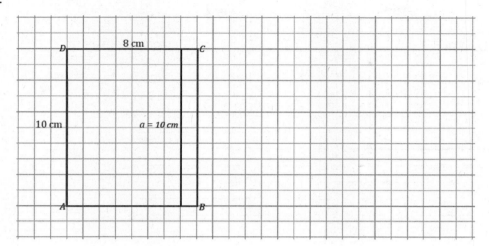

b.

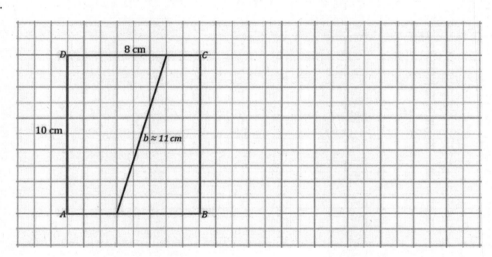

c.

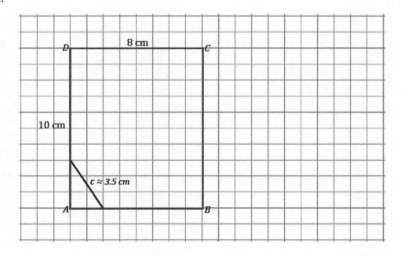

©2015 Great Minds. eureka-math.org
G7-M6-SE-B3-1.3.1-01.2016

d.

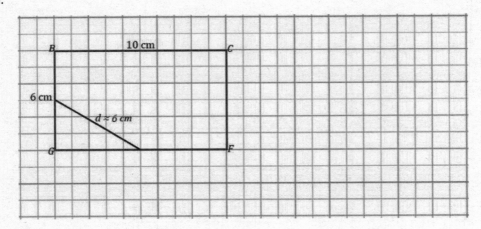

e.

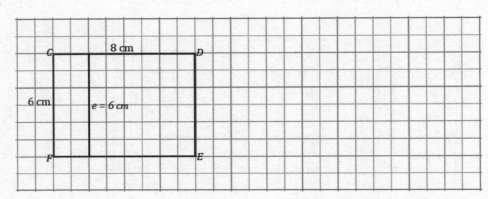

f.

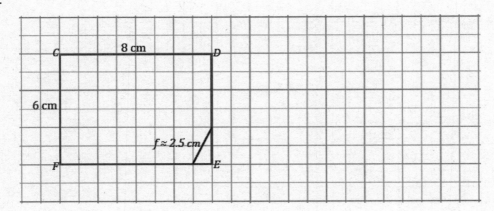

EUREKA
MATH™

g.

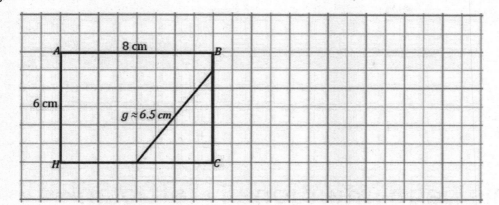

EUREKA
MATH™

©2015 Great Minds. eureka-math.org
G7-M6-SE-B3-1.3.1-01.2016

This page intentionally left blank

Lesson 17: Slicing a Right Rectangular Pyramid with a Plane

Classwork

Opening

RECTANGULAR PYRAMID: Given a rectangular region B in a plane E, and a point V not in E, the *rectangular pyramid with base B and vertex V* is the collection of all segments VP for any point P in B. It can be shown that the planar region defined by a side of the base B and the vertex V is a triangular region called a *lateral face*.

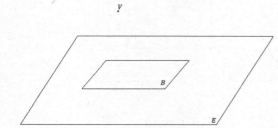

A rectangular region B in a plane E and a point V not in E

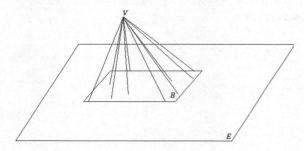

The rectangular pyramid is determined by the collection of all segments VP for any point P in B; here $\overline{VP}$ is shown for a total of 10 points.

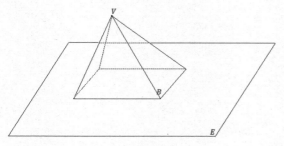

The rectangular pyramid is a solid once the collection of all segments VP for any point P in B are taken. The pyramid has a total of five faces: four lateral faces and a base.

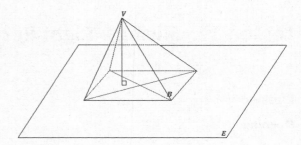

If the vertex lies on the line perpendicular to the base at its center (i.e., the intersection of the rectangle's diagonals), the pyramid is called a *right rectangular pyramid*. The example of the rectangular pyramid above is not a right rectangular pyramid, as evidenced in this image. The perpendicular from V does not meet at the intersection of the diagonals of the rectangular base B.

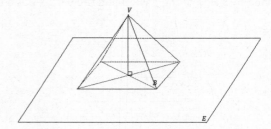

The following is an example of a right rectangular pyramid. The opposite lateral faces are identical isosceles triangles.

Example 1

Use the models you built to assist in a sketch of a pyramid. Though you are sketching from a model that is opaque, use dotted lines to represent the edges that cannot be seen from your perspective.

EUREKA
MATH™

Example 2

Sketch a right rectangular pyramid from three vantage points: (1) from directly over the vertex, (2) from facing straight on to a lateral face, and (3) from the bottom of the pyramid. Explain how each drawing shows each view of the pyramid.

Example 3

Assume the following figure is a top-down view of a rectangular pyramid. Make a reasonable sketch of any two adjacent lateral faces. What measurements must be the same between the two lateral faces? Mark the equal measurement. Justify your reasoning for your choice of equal measurements.

Example 4

a. A slicing plane passes through segment a parallel to base B of the right rectangular pyramid below. Sketch what the slice will look like into the figure. Then sketch the resulting slice as a two-dimensional figure. Students may choose how to represent the slice (e.g., drawing a 2D or 3D sketch or describing the slice in words).

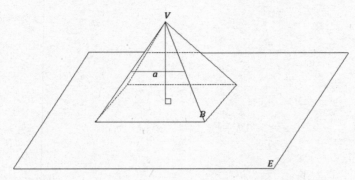

b. What shape does the slice make? What is the relationship between the slice and the rectangular base of the pyramid?

EUREKA
MATH™

Example 5

A slice is to be made along segment a perpendicular to base B of the right rectangular pyramid below.

a. Which of the following figures shows the correct slice? Justify why each of the following figures is or is not a correct diagram of the slice.

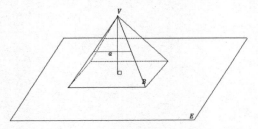

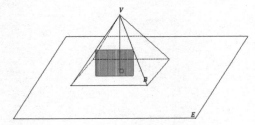

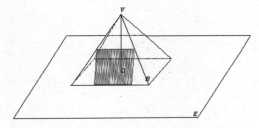

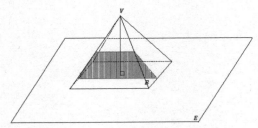

b. A slice is taken through the vertex of the pyramid perpendicular to the base. Sketch what the slice will look like into the figure. Then, sketch the resulting slice itself as a two-dimensional figure.

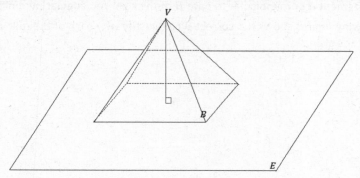

EUREKA
MATH™

Lesson Summary

- A rectangular pyramid differs from a right rectangular pyramid because the vertex of a right rectangular pyramid lies on the line perpendicular to the base at its center whereas a pyramid that is not a right rectangular pyramid will have a vertex that is not on the line perpendicular to the base at its center.
- Slices made parallel to the base of a right rectangular pyramid are scale drawings of the rectangular base of the pyramid.

Problem Set

A side view of a right rectangular pyramid is given. The line segments lie in the lateral faces.

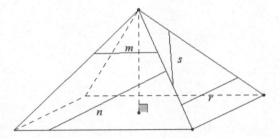

a. For segments n, s, and r, sketch the resulting slice from slicing the right rectangular pyramid with a slicing plane that contains the line segment and is perpendicular to the base.

b. For segment m, sketch the resulting slice from slicing the right rectangular pyramid with a slicing plane that contains the segment and is parallel to the base.

Note: To challenge yourself, you can try drawing the slice into the pyramid.

c. A top view of a right rectangular pyramid is given. The line segments lie in the base face. For each line segment, sketch the slice that results from slicing the right rectangular pyramid with a plane that contains the line segment and is perpendicular to the base.

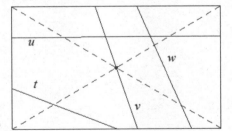

This page intentionally left blank

Lesson 18: Slicing on an Angle

Classwork

Example 1

With your group, discuss whether a right rectangular prism can be sliced at an angle so that the resulting slice looks like the figure in Figure 1. If it is possible, draw an example of such a slice into the following prism.

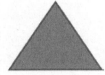

Figure 1

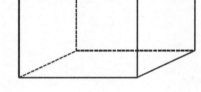

Exercise 1

    a.   With your group, discuss how to slice a right rectangular prism so that the resulting slice looks like the figure in Figure 2. Justify your reasoning.

Figure 2

    b.   With your group, discuss how to slice a right rectangular prism so that the resulting slice looks like the figure in Figure 3. Justify your reasoning.

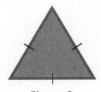

Figure 3

EUREKA
MATH™

Example 2

With your group, discuss whether a right rectangular prism can be sliced at an angle so that the resulting slice looks like the figure in Figure 4. If it is possible, draw an example of such a slice into the following prism.

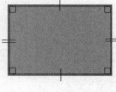

Figure 4

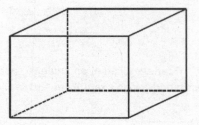

Exercise 2

In Example 2, we discovered how to slice a right rectangular prism to makes the shapes of a rectangle and a parallelogram. Are there other ways to slice a right rectangular prism that result in other quadrilateral-shaped slices?

Example 3

a. If slicing a plane through a right rectangular prism so that the slice meets the three faces of the prism, the resulting slice is in the shape of a triangle; if the slice meets four faces, the resulting slice is in the shape of a quadrilateral. Is it possible to slice the prism in a way that the region formed is a pentagon (as in Figure 5)? A hexagon (as in Figure 6)? An octagon (as in Figure 7)?

Figure 5

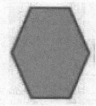

Figure 6

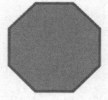

Figure 7

b.   Draw an example of a slice in a pentagon shape and a slice in a hexagon shape.

Example 4

a.   With your group, discuss whether a right rectangular pyramid can be sliced at an angle so that
     the resulting slice looks like the figure in Figure 8. If it is possible, draw an example of such a
     slice into the following pyramid.

Figure 8

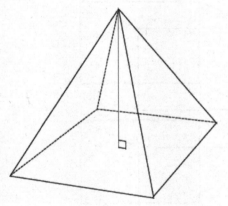

b.   With your group, discuss whether a right rectangular pyramid can be sliced at an angle so that
     the resulting slice looks like the figure in Figure 9. If it is possible, draw an example of such a
     slice into the pyramid above.

Figure 9

Lesson Summary

- Slices made at an angle are neither parallel nor perpendicular to a base.
- There cannot be more sides to the polygonal region of a slice than there are faces of the solid.

Problem Set

1. Draw a slice into the right rectangular prism at an angle in the form of the provided shape, and draw each slice as a 2D shape.

Slice made in the prism   **Slice as a 2D shape**

a.   A triangle

b.   A quadrilateral

c.   A pentagon

d.   A hexagon

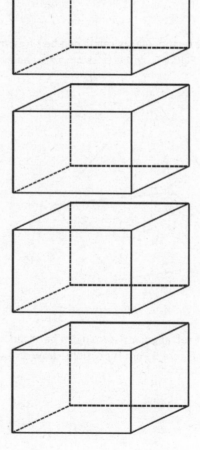

2. Draw slices at an angle in the form of each given shape into each right rectangular pyramid, and draw each slice as a 2D shape.

Slice made in the pyramid   **Slice as a 2D shape**

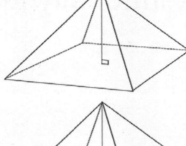

a. A triangle

b. A quadrilateral

c. A pentagon

3. Why is it not possible to draw a slice in the shape of a hexagon for a right rectangular pyramid?

4. If the slicing plane meets every face of a right rectangular prism, then the slice is a hexagonal region. What can you say about opposite sides of the hexagon?

5. Draw a right rectangular prism so that rectangles $ABCD$ and $A'B'C'D'$ are base faces. The line segments AA', BB', CC', and DD' are edges of the lateral faces.
    a. A slicing plane meets the prism so that vertices A, B, C, and D lie on one side of the plane, and vertices A', B', C', and D' lie on the other side. Based on the slice's position, what other information can be concluded about the slice?
    b. A slicing plane meets the prism so that vertices A, B, C, and B' are on one side of the plane, and vertices A', D', C', and D are on the other side. What other information can be concluded about the slice based on its position?

This page intentionally left blank

Lesson 19: Understanding Three-Dimensional Figures

Classwork

Example 1

If slices parallel to the tabletop (with height a whole number of units from the tabletop) were taken of this figure, then what would each slice look like?

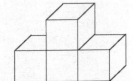

Example 2

If slices parallel to the tabletop were taken of this figure, then what would each slice look like?

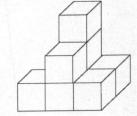

Exercise 1

Based on the level slices you determined in Example 2, how many unit cubes are in the figure?

Exercise 2

a.   If slices parallel to the tabletop were taken of this figure, then what would each slice look like?

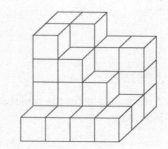

b.   Given the level slices in the figure, how many unit cubes are in the figure?

Example 3

Given the level slices in the figure, how many unit cubes are in the figure?

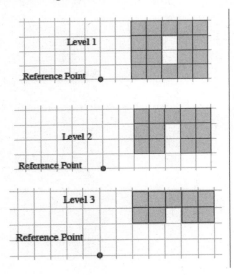
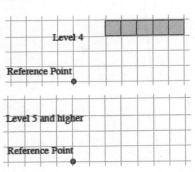

©2015 Great Minds. eureka-math.org
G7-M6-SE-B3-1.3.1-01.2016

Exercise 3

Sketch your own three-dimensional figure made from cubes and the slices of your figure. Explain how the slices relate to the figure.

Lesson 19:   Understanding Three-Dimensional Figures

Lesson Summary

We can examine the horizontal whole-unit scales to look at three-dimensional figures. These slices allow a way to count the number of unit cubes in the figure, which is useful when the figure is layered in a way so that many cubes are hidden from view.

Problem Set

In the given three-dimensional figures, unit cubes are stacked exactly on top of each other on a tabletop. Each block is either visible or below a visible block.

1.

    a.   The following three-dimensional figure is built on a tabletop. If slices parallel to the tabletop are taken of this figure, then what would each slice look like?

    b.   Given the level slices in the figure, how many cubes are in the figure?

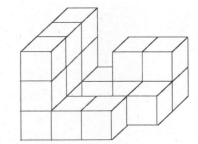

2.

    a.   The following three-dimensional figure is built on a tabletop. If slices parallel to the tabletop are taken of this figure, then what would each slice look like?

    b.   Given the level slices in the figure, how many cubes are in the figure?

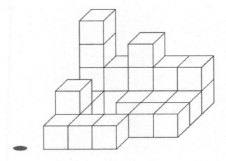

3.

    a.   The following three-dimensional figure is built on a tabletop. If slices parallel to the tabletop are taken of this figure, then what would each slice look like?

    b.   Given the level slices in the figure, how many cubes are in the figure?

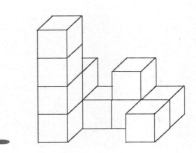

4. John says that we should be including the Level 0 slice when mapping slices. Naya disagrees, saying it is correct to start counting cubes from the Level 1 slice. Who is right?

5. Draw a three-dimensional figure made from cubes so that each successive layer farther away from the tabletop has one less cube than the layer below it. Use a minimum of three layers. Then draw the slices, and explain the connection between the two.

Lesson 20: Real-World Area Problems

Classwork

Opening Exercise

Find the area of each shape based on the provided measurements. Explain how you found each area.

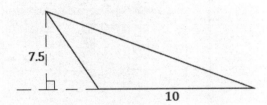

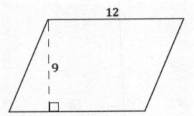

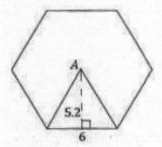

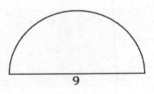

Example 1

A landscape company wants to plant lawn seed. A 20 lb. bag of lawn seed will cover up to 420 sq. ft. of grass and costs $49.98 plus the 8% sales tax. A scale drawing of a rectangular yard is given. The length of the longest side is 100 ft. The house, driveway, sidewalk, garden areas, and utility pad are shaded. The unshaded area has been prepared for planting grass. How many 20 lb. bags of lawn seed should be ordered, and what is the cost?

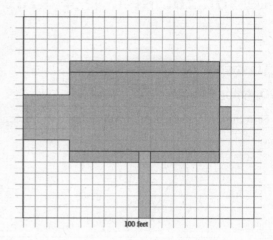

100 feet

Exercise 1

A landscape contractor looks at a scale drawing of a yard and estimates that the area of the home and garage is the same as the area of a rectangle that is 100 ft.× 35 ft. The contractor comes up with 5,500 ft². How close is this estimate?

Example 2

Ten dartboard targets are being painted as shown in the following figure. The radius of the smallest circle is 3 in., and each successive larger circle is 3 in. more in radius than the circle before it. A can of red paint and a can of white paint is purchased to paint the target. Each 8 oz. can of paint covers 16 ft². Is there enough paint of each color to create all ten targets?

Lesson Summary

- One strategy to use when solving area problems with real-world context is to decompose drawings into familiar polygons and circular regions while identifying all relevant measurements.

- Since the area problems involve real-world context, it is important to pay attention to the units needed in each response.

Problem Set

1. A farmer has four pieces of unfenced land as shown to the right in the scale drawing where the dimensions of one side are given. The farmer trades all of the land and $10,000 for 8 acres of similar land that is fenced. If one acre is equal to 43,560 ft², how much per square foot for the extra land did the farmer pay rounded to the nearest cent?

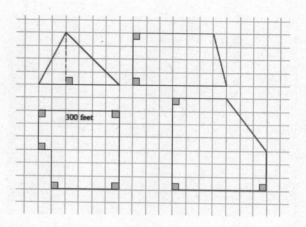

2. An ordinance was passed that required farmers to put a fence around their property. The least expensive fences cost $10 for each foot. Did the farmer save money by moving the farm?

3. A stop sign is an octagon (i.e., a polygon with eight sides) with eight equal sides and eight equal angles. The dimensions of the octagon are given. One side of the stop sign is to be painted red. If Timmy has enough paint to cover 500 ft², can he paint 100 stop signs? Explain your answer.

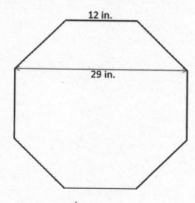

4. The Smith family is renovating a few aspects of their home. The following diagram is of a new kitchen countertop. Approximately how many square feet of counter space is there?

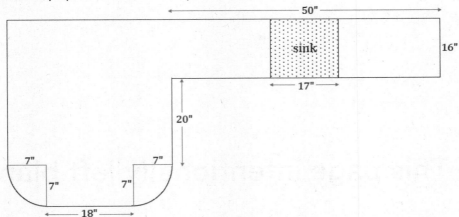

5. In addition to the kitchen renovation, the Smiths are laying down new carpet. Everything but closets, bathrooms, and the kitchen will have new carpet. How much carpeting must be purchased for the home?

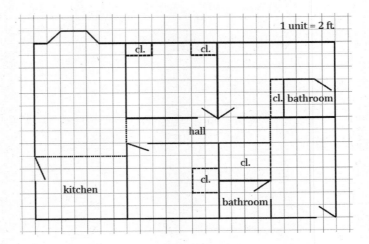

6. Jamie wants to wrap a rectangular sheet of paper completely around cans that are $8\frac{1}{2}$ in. high and 4 in. in diameter. She can buy a roll of paper that is $8\frac{1}{2}$ in. wide and 60 ft. long. How many cans will this much paper wrap?

This page intentionally left blank

Lesson 21: Mathematical Area Problems

Classwork

Opening Exercise

Patty is interested in expanding her backyard garden. Currently, the garden plot has a length of 4 ft. and a width of 3 ft.

a.   What is the current area of the garden?

Patty plans on extending the length of the plot by 3 ft. and the width by 2 ft.

b.   What will the new dimensions of the garden be? What will the new area of the garden be?

c.   Draw a diagram that shows the change in dimension and area of Patty's garden as she expands it. The diagram should show the original garden as well as the expanded garden.

d. Based on your diagram, can the area of the garden be found in a way other than by multiplying the length by the width?

e. Based on your diagram, how would the area of the original garden change if only the length increased by 3 ft.? By how much would the area increase?

f. How would the area of the original garden change if only the width increased by 2 ft.? By how much would the area increase?

g. Complete the following table with the numeric expression, area, and increase in area for each change in the dimensions of the garden.

Dimensions of the Garden	Numeric Expression for the Area of the Garden	Area of the Garden	Increase in Area of the Garden
The original garden with length of 4 ft. and width of 3 ft.			
The original garden with length extended by 3 ft. and width extended by 2 ft.			
The original garden with only the length extended by 3 ft.			
The original garden with only the width extended by 2 ft.			

h. Will the increase in both the length and width by 3 ft. and 2 ft., respectively, mean that the original area will increase strictly by the areas found in parts (e) and (f)? If the area is increasing by more than the areas found in parts (e) and (f), explain what accounts for the additional increase.

Example 1

Examine the change in dimension and area of the following square as it increases by 2 units from a side length of 4 units to a new side length of 6 units. Observe the way the area is calculated for the new square. The lengths are given in units, and the areas of the rectangles and squares are given in units squared.

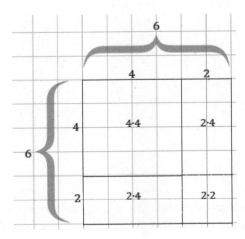

a. Based on the example above, draw a diagram for a square with a side length of 3 units that is increasing by 2 units. Show the area calculation for the larger square in the same way as in the example.

b.   Draw a diagram for a square with a side length of 5 units that is increased by 3 units. Show the area calculation for the larger square in the same way as in the example.

c.   Generalize the pattern for the area calculation of a square that has an increase in dimension. Let the length of the original square be a units and the increase in length be b units. Use the diagram below to guide your work.

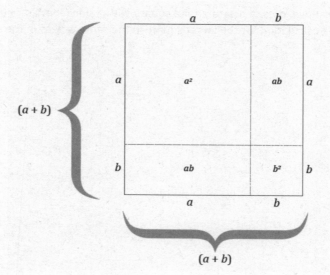

Example 2

Bobby draws a square that is 10 units by 10 units. He increases the length by x units and the width by 2 units.

    a.   Draw a diagram that models this scenario.

    b.   Assume the area of the large rectangle is 156 units^2. Find the value of x.

Example 3

The dimensions of a square with a side length of x units are increased. In this figure, the indicated lengths are given in units, and the indicated areas are given in units^2.

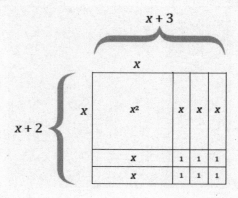

a.   What are the dimensions of the large rectangle in the figure?

b.   Use the expressions in your response from part (a) to write an equation for the area of the large rectangle, where A represents area.

c.   Use the areas of the sections within the diagram to express the area of the large rectangle.

d.   What can be concluded from parts (b) and (c)?

e.   Explain how the expressions $(x + 2)(x + 3)$ and $x^2 + 3x + 2x + 6$ differ within the context of the area of the figure.

EUREKA
MATH™

Lesson Summary

- The properties of area are limited to positive numbers for lengths and areas.
- The properties of area do support why the properties of operations are true.

Problem Set

1. A square with a side length of a units is decreased by b units in both length and width.

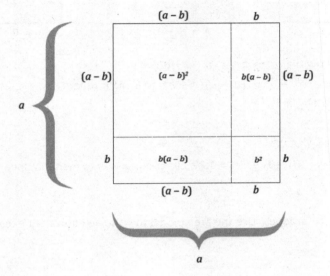

Use the diagram to express $(a - b)^2$ in terms of the other a^2, ab, and b^2 by filling in the blanks below:

$(a - b)^2 = a^2 - b(a - b) - b(a - b) - b^2$

$\qquad = a^2 - \underline{\quad} + \underline{\quad} - \underline{\quad} + \underline{\quad} - b^2$

$\qquad = a^2 - 2ab + \underline{\quad} - b^2$

$\qquad = \underline{\hspace{3cm}}$

2. In Example 3, part (c), we generalized that $(a + b)^2 = a^2 + 2ab + b^2$. Use these results to evaluate the following expressions by writing $1,001 = 1,000 + 1$.

   a. Evaluate 101^2.

   b. Evaluate $1,001^2$.

   c. Evaluate 21^2.

3. Use the results of Problem 1 to evaluate 999^2 by writing $999 = 1,000 - 1$.

4. The figures below show that $8^2 - 5^2$ is equal to $(8-5)(8+5)$.

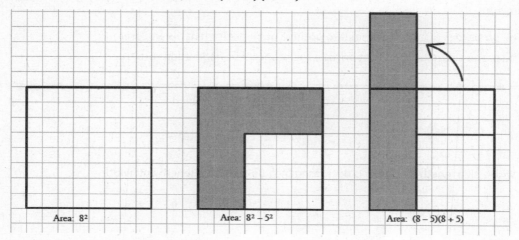

Area: 8^2 Area: $8^2 - 5^2$ Area: $(8-5)(8+5)$

a. Create a drawing to show that $a^2 - b^2 = (a-b)(a+b)$.

b. Use the result in part (a), $a^2 - b^2 = (a-b)(a+b)$, to explain why:

    i. $35^2 - 5^2 = (30)(40)$.

    ii. $21^2 - 18^2 = (3)(39)$.

    iii. $104^2 - 63^2 = (41)(167)$.

c. Use the fact that $35^2 = (30)(40) + 5^2 = 1,225$ to create a way to mentally square any two-digit number ending in 5.

5. Create an area model for each product. Use the area model to write an equivalent expression that represents the area.

a. $(x+1)(x+4) =$

b. $(x+5)(x+2) =$

c. Based on the context of the area model, how do the expressions provided in parts (a) and (b) differ from the equivalent expression answers you found for each?

6. Use the distributive property to multiply the following expressions.

a. $(2+6)(2+4)$

b. $(x+6)(x+4)$; draw a figure that models this multiplication problem.

c. $(10+7)(10+7)$

d. $(a+7)(a+7)$

e. $(5-3)(5+3)$

f. $(x-3)(x+3)$

Lesson 22: Area Problems with Circular Regions

Classwork

Example 1

a. The circle to the right has a diameter of 12 cm. Calculate the area of the shaded region.

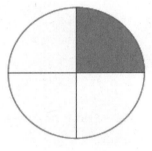

b. Sasha, Barry, and Kyra wrote three different expressions for the area of the shaded region. Describe what each student was thinking about the problem based on his or her expression.

Sasha's expression: $\frac{1}{4}\pi(6^2)$

Barry's expression: $\pi(6^2) - \frac{3}{4}\pi(6^2)$

Kyra's expression: $\frac{1}{2}\left(\frac{1}{2}\pi(6^2)\right)$

Exercise 1

a.   Find the area of the shaded region of the circle to the right.

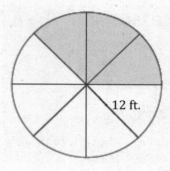

12 ft.

b.   Explain how the expression you used represents the area of the shaded region.

Exercise 2

Calculate the area of the figure below that consists of a rectangle and two quarter circles, each with the same radius. Leave your answer in terms of pi.

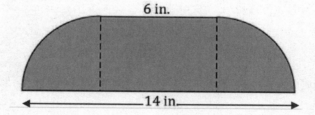

6 in.

14 in.

Example 2

The square in this figure has a side length of 14 inches. The radius of the quarter circle is 7 inches.

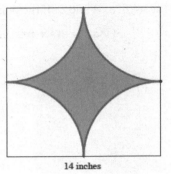

14 inches

    a.   Estimate the shaded area.

    b.   What is the exact area of the shaded region?

    c.   What is the approximate area using $\pi \approx \frac{22}{7}$?

Exercise 3

The vertices A and B of rectangle $ABCD$ are centers of circles each with a radius of 5 inches.

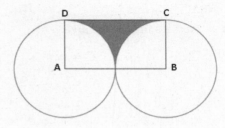

a. Find the exact area of the shaded region.

b. Find the approximate area using $\pi \approx \frac{22}{7}$.

c. Find the area to the nearest hundredth using the π key on your calculator.

EUREKA
MATH™

Exercise 4

The diameter of the circle is 12 in. Write and explain a numerical expression that represents the area of the shaded region.

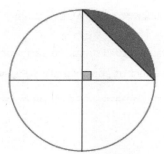

Lesson Summary

To calculate composite figures with circular regions:

- Identify relevant geometric areas (such as rectangles or squares) that are part of a figure with a circular region.

- Determine which areas should be subtracted or added based on their positions in the diagram.

- Answer the question, noting if the exact or approximate area is to be found.

Problem Set

1. A circle with center O has an area of 96 in². Find the area of the shaded region.

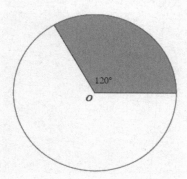

120°

O

Peyton's Solution

$$A = \frac{1}{3}(96 \text{ in}^2) = 32 \text{ in}^2$$

Monte's Solution

$$A = \frac{96}{120} \text{ in}^2 = 0.8 \text{ in}^2$$

Which person solved the problem correctly? Explain your reasoning.

2. The following region is bounded by the arcs of two quarter circles, each with a radius of 4 cm, and by line segments 6 cm in length. The region on the right shows a rectangle with dimensions 4 cm by 6 cm. Show that both shaded regions have equal areas.

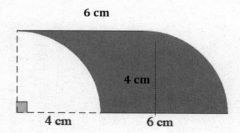

6 cm

4 cm

4 cm

6 cm

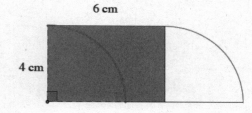

6 cm

4 cm

3. A square is inscribed in a paper disc (i.e., a circular piece of paper) with a radius of 8 cm. The paper disc is red on the front and white on the back. Two edges of the circle are folded over. Write and explain a numerical expression that represents the area of the figure. Then, find the area of the figure.

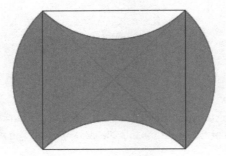

4. The diameters of four half circles are sides of a square with a side length of 7 cm.

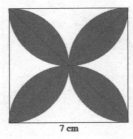

7 cm

a. Find the exact area of the shaded region.

b. Find the approximate area using $\pi \approx \frac{22}{7}$.

c. Find the area using the π button on your calculator and rounding to the nearest thousandth.

5. A square with a side length of 14 inches is shown below, along with a quarter circle (with a side of the square as its radius) and two half circles (with diameters that are sides of the square). Write and explain a numerical expression that represents the area of the figure.

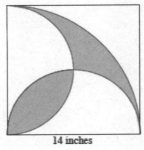

14 inches

6. Three circles have centers on segment AB. The diameters of the circles are in the ratio $3:2:1$. If the area of the largest circle is 36 ft^2, find the area inside the largest circle but outside the smaller two circles.

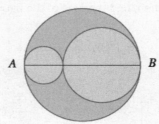

7. A square with a side length of 4 ft. is shown, along with a diagonal, a quarter circle (with a side of the square as its radius), and a half circle (with a side of the square as its diameter). Find the exact, combined area of regions I and II.

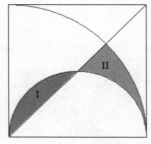

EUREKA
MATH™

Lesson 23: Surface Area

Classwork

Opening Exercise

Calculate the surface area of the square pyramid.

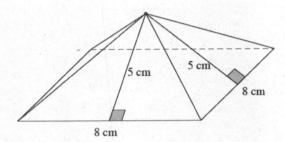

Example 1

a. Calculate the surface area of the rectangular prism.

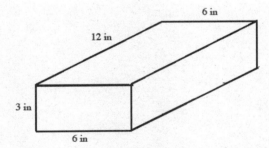

©2015 Great Minds. eureka-math.org
G7-M6-SE-B3-1.3.1-01.2016

b.   Imagine that a piece of the rectangular prism is removed. Determine the surface area of both pieces.

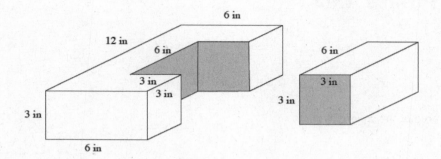

c.   How is the surface area in part (a) related to the surface area in part (b)?

Exercises

Determine the surface area of the right prisms.

1.

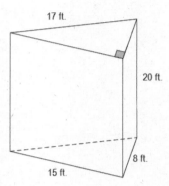

2.

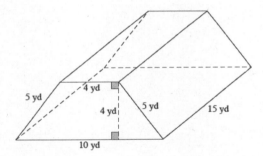

3.

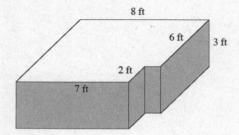

4.

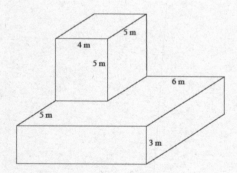

Lesson 23:   Surface Area

©2015 Great Minds. eureka-math.org
G7-M6-SE-B3-1.3.1-01.2016

5.

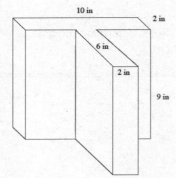

Lesson Summary

To determine the surface area of right prisms that are composite figures or missing sections, determine the area of each lateral face and the two base faces, and then add the areas of all the faces together.

Problem Set

Determine the surface area of the figures.

1.

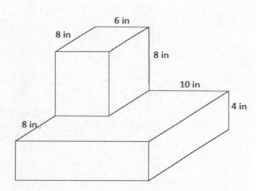

2.

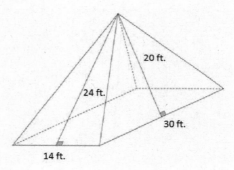

3.

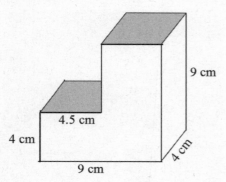

4.

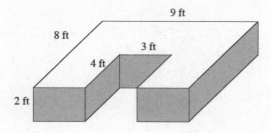

5.

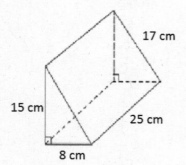

Lesson 24: Surface Area

Classwork

Example 1

Determine the surface area of the image.

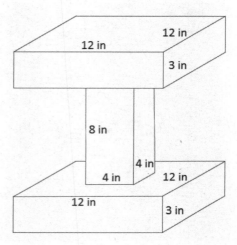

Example 2

a.   Determine the surface area of the cube.

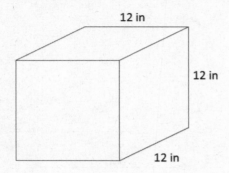

b.   A square hole with a side length of 4 inches is cut through the cube. Determine the new surface area.

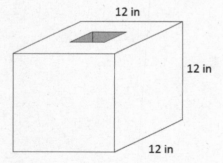

Example 3

A right rectangular pyramid has a square base with a side length of 10 inches. The surface area of the pyramid is 260 in². Find the height of the four lateral triangular faces.

Exercises

Determine the surface area of each figure. Assume all faces are rectangles unless it is indicated otherwise.

1.

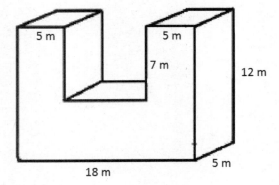

2. In addition to your calculation, explain how the surface area of the following figure was determined.

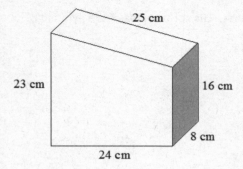

3.

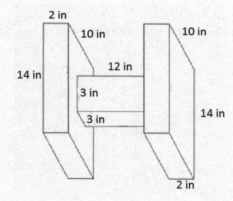

4. In addition to your calculation, explain how the surface area was determined.

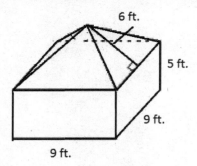

5. A hexagonal prism has the following base and has a height of 8 units. Determine the surface area of the prism.

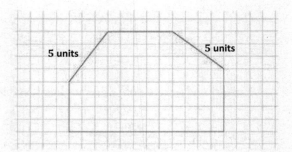

6. Determine the surface area of each figure.

    a.

    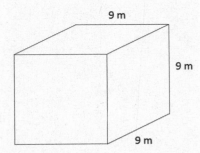

    b. A cube with a square hole with 3 m side lengths has been cut through the cube.

    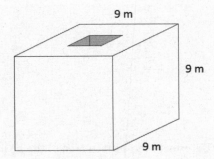

    c. A second square hole with 3 m side lengths has been cut through the cube.

    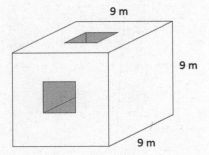

7.   The figure below shows 28 cubes with an edge length of 1 unit. Determine the surface area.

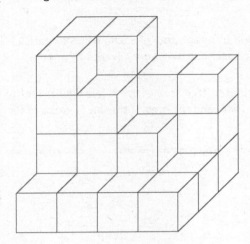

8.   The base rectangle of a right rectangular prism is 4 ft.× 6 ft. The surface area is 288 ft². Find the height.
     Let h be the height in feet.

Lesson Summary

- To calculate the surface area of a composite figure, determine the surface area of each prism separately, and add them together. From the sum, subtract the area of the sections that were covered by another prism.

- To calculate the surface area with a missing section, find the total surface area of the whole figure. From the total surface area, subtract the area of the missing parts. Then, add the area of the lateral faces of the cutout prism.

Problem Set

Determine the surface area of each figure.

1. In addition to the calculation of the surface area, describe how you found the surface area.

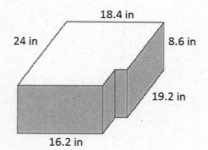

2.

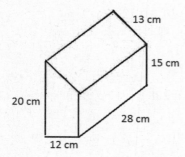

3.

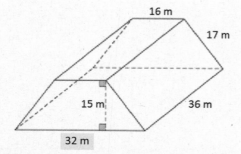

EUREKA
MATH™

4. Determine the surface area after two square holes with a side length of 2 m are cut through the solid figure composed of two rectangular prisms.

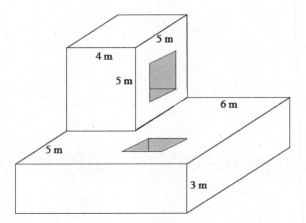

5. The base of a right prism is shown below. Determine the surface area if the height of the prism is 10 cm. Explain how you determined the surface area.

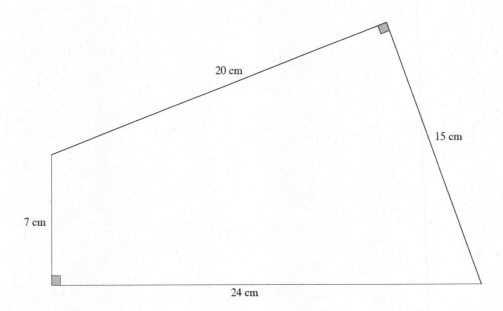

This page intentionally left blank

Lesson 25: Volume of Right Prisms

Classwork

Opening Exercise

Take your copy of the following figure, and cut it into four pieces along the dotted lines. (The vertical line is the altitude, and the horizontal line joins the midpoints of the two sides of the triangle.)

Arrange the four pieces so that they fit together to form a rectangle.

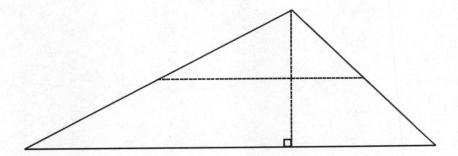

If a prism were formed out of each shape, the original triangle, and your newly rearranged rectangle, and both prisms had the same height, would they have the same volume? Discuss with a partner.

Exercise 1

a. Show that the following figures have equal volumes.

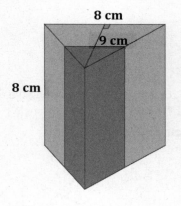 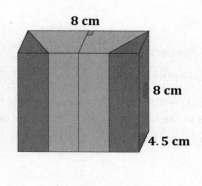

b. How can it be shown that the prisms will have equal volumes without completing the entire calculation?

Example 1

Calculate the volume of the following prism.

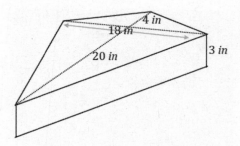

EUREKA
MATH™

Example 2

A container is shaped like a right pentagonal prism with an open top. When a cubic foot of water is dumped into the container, the depth of the water is 8 inches. Find the area of the pentagonal base.

Example 3

Two containers are shaped like right triangular prisms, each with the same height. The base area of the larger container is 200% more than the base area of the smaller container. How many times must the smaller container be filled with water and poured into the larger container in order to fill the larger container?

Exercise 2

Two aquariums are shaped like right rectangular prisms. The ratio of the dimensions of the larger aquarium to the dimensions of the smaller aquarium is $3:2$.

Addie says the larger aquarium holds 50% more water than the smaller aquarium.

Berry says that the larger aquarium holds 150% more water.

Cathy says that the larger aquarium holds over 200% more water.

Are any of the girls correct? Explain your reasoning.

©2015 Great Minds. eureka-math.org
G7-M6-SE-B3-1.3.1-01.2016

Lesson Summary

- The formula for the volume of a prism is $V = Bh$, where B is the area of the base of the prism and h is the height of the prism.

- A base that is neither a rectangle nor a triangle must be decomposed into rectangles and triangles in order to find the area of the base.

Problem Set

1. The pieces in Figure 1 are rearranged and put together to form Figure 2.

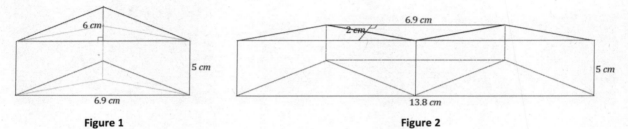

Figure 1   **Figure 2**

   a. Use the information in Figure 1 to determine the volume of the prism.

   b. Use the information in Figure 2 to determine the volume of the prism.

   c. If we were not told that the pieces of Figure 1 were rearranged to create Figure 2, would it be possible to determine whether the volumes of the prisms were equal without completing the entire calculation for each?

2. Two right prism containers each hold 75 gallons of water. The height of the first container is 20 inches. The of the second container is 30 inches. If the area of the base in the first container is 6 ft^2, find the area of the base in the second container. Explain your reasoning.

3. Two containers are shaped like right rectangular prisms. Each has the same height, but the base of the larger container is 50% more in each direction. If the smaller container holds 8 gallons when full, how many gallons does the larger container hold? Explain your reasoning.

4. A right prism container with the base area of 4 ft^2 and height of 5 ft. is filled with water until it is 3 ft. deep. If a solid cube with edge length 1 ft. is dropped to the bottom of the container, how much will the water rise?

5. A right prism container with a base area of 10 ft^2 and height 9 ft. is filled with water until it is 6 ft. deep. A large boulder is dropped to the bottom of the container, and the water rises to the top, completely submerging the boulder without causing overflow. Find the volume of the boulder.

6. A right prism container with a base area of 8 ft^2 and height 6 ft. is filled with water until it is 5 ft. deep. A solid cube is dropped to the bottom of the container, and the water rises to the top. Find the length of the cube.

7. A rectangular swimming pool is 30 feet wide and 50 feet long. The pool is 3 feet deep at one end, and 10 feet deep at the other.

   a.   Sketch the swimming pool as a right prism.

   b.   What kind of right prism is the swimming pool?

   c.   What is the volume of the swimming pool in cubic feet?

   d.   How many gallons will the swimming pool hold if each cubic feet of water is about 7.5 gallons?

8. A milliliter (mL) has a volume of 1 cm^3. A 250 mL measuring cup is filled to 200 mL. A small stone is placed in the measuring cup. The stone is completely submerged, and the water level rises to 250 mL.

   a.   What is the volume of the stone in cm^3?

   b.   Describe a right rectangular prism that has the same volume as the stone.

Opening Exercise

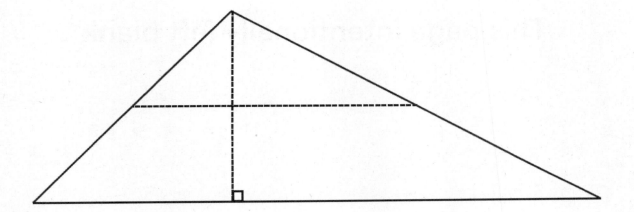

This page intentionally left blank

Lesson 26: Volume of Composite Three-Dimensional Objects

Classwork

Example 1

Find the volume of the following three-dimensional object composed of two right rectangular prisms.

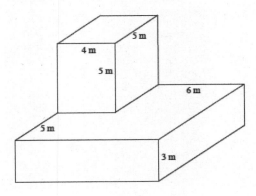

Exercise 1

Find the volume of the following three-dimensional figure composed of two right rectangular prisms.

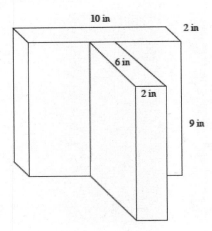

EUREKA MATH™

©2015 Great Minds. eureka-math.org
G7-M6-SE-B3-1.3.1-01.2016

Exercise 2

The right trapezoidal prism is composed of a right rectangular prism joined with a right triangular prism. Find the volume of the right trapezoidal prism shown in the diagram using two different strategies.

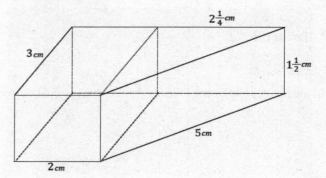

Lesson 26:   Volume of Composite Three-Dimensional Objects

Example 2

Find the volume of the right prism shown in the diagram whose base is the region between two right triangles. Use two different strategies.

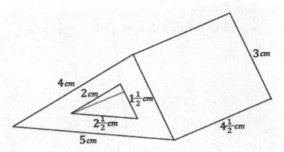

Example 3

A box with a length of 2 ft., a width of 1.5 ft., and a height of 1.25 ft. contains fragile electronic equipment that is packed inside a larger box with three inches of styrofoam cushioning material on each side (above, below, left side, right side, front, and back).

a.   Give the dimensions of the larger box.

b.   Design styrofoam right rectangular prisms that could be placed around the box to provide the cushioning (i.e., give the dimensions and how many of each size are needed).

c.   Find the volume of the styrofoam cushioning material by adding the volumes of the right rectangular prisms in the previous question.

d.   Find the volume of the styrofoam cushioning material by computing the difference between the volume of the larger box and the volume of the smaller box.

Lesson Summary

To find the volume of a three-dimensional composite object, two or more distinct volumes must be added together (if they are joined together) or subtracted from each other (if one is a missing section of the other). There are two strategies to find the volume of a prism:

- Find the area of the base and then multiply times the prism's height.
- Decompose the prism into two or more smaller prisms of the same height and add the volumes of those smaller prisms.

Problem Set

1. Find the volume of the three-dimensional object composed of right rectangular prisms.

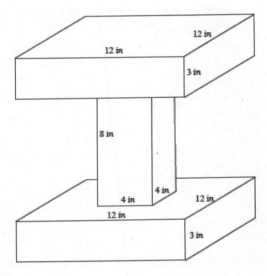

2. A smaller cube is stacked on top of a larger cube. An edge of the smaller cube measures $\frac{1}{2}$ cm in length, while the larger cube has an edge length three times as long. What is the total volume of the object?

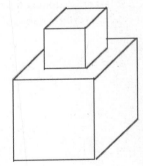

3. Two students are finding the volume of a prism with a rhombus base but are provided different information regarding the prism. One student receives Figure 1, while the other receives Figure 2.

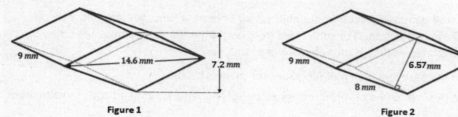

Figure 1   Figure 2

a. Find the expression that represents the volume in each case; show that the volumes are equal.

b. How does each calculation differ in the context of how the prism is viewed?

4. Find the volume of wood needed to construct the following side table composed of right rectangular prisms.

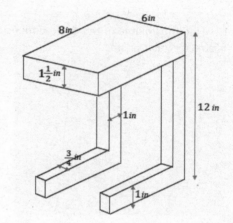

5. A plastic die (singular for dice) for a game has an edge length of 1.5 cm. Each face of the cube has the number of cubic cutouts as its marker is supposed to indicate (i.e., the face marked 3 has 3 cutouts). What is the volume of the die?

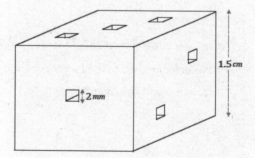

6. A wooden cube with an edge length of 6 inches has square holes (holes in the shape of right rectangular prisms) cut through the centers of each of the three sides as shown in the figure. Find the volume of the resulting solid if the square for the holes has an edge length of 1 inch.

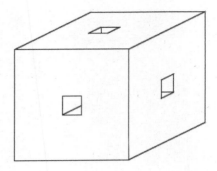

7. A right rectangular prism has each of its dimensions (length, width, and height) increased by 50%. By what percent is its volume increased?

8. A solid is created by putting together right rectangular prisms. If each of the side lengths is increase by 40%, by what percent is the volume increased?

This page intentionally left blank

Lesson 27: Real-World Volume Problems

Classwork

Example 1

A swimming pool holds 10,000 ft^3 of water when filled. Jon and Anne want to fill the pool with a garden hose. The garden hose can fill a five-gallon bucket in 30 seconds. If each cubic foot is about 7.5 gallons, find the flow rate of the garden hose in gallons per minute and in cubic feet per minute. About how long will it take to fill the pool with a garden hose? If the hose is turned on Monday morning at 8:00 a.m., approximately when will the pool be filled?

Example 2

A square pipe (a rectangular prism-shaped pipe) with inside dimensions of 2 in. × 2 in. has water flowing through it at a flow speed of $3\frac{\text{ft}}{\text{s}}$. The water flows into a pool in the shape of a right triangular prism, with a base in the shape of a right isosceles triangle and with legs that are each 5 feet in length. How long will it take for the water to reach a depth of 4 feet?

Exercise 1

A park fountain is about to be turned on in the spring after having been off all winter long. The fountain flows out of the top level and into the bottom level until both are full, at which point the water is just recycled from top to bottom through an internal pipe. The outer wall of the top level, a right square prism, is five feet in length; the thickness of the stone between outer and inner wall is 1 ft.; and the depth is 1 ft. The bottom level, also a right square prism, has an outer wall that is 11 ft. long with a 2 ft. thickness between the outer and inner wall and a depth of 2 ft. Water flows through a 3 in. × 3 in. square pipe into the top level of the fountain at a flow speed of $4\frac{\text{ft}}{\text{s}}$. Approximately how long will it take for both levels of the fountain to fill completely?

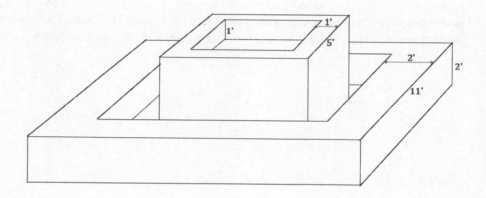

Exercise 2

A decorative bathroom faucet has a 3 in. × 3 in. square pipe that flows into a basin in the shape of an isosceles trapezoid prism like the one shown in the diagram. If it takes one minute and twenty seconds to fill the basin completely, what is the approximate speed of water flowing from the faucet in feet per second?

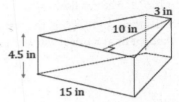

Lesson Summary

The formulas $V = Bh$ and $V = rt$, where r is flow rate, can be used to solve real-world volume problems involving flow speed and flow rate. For example, water flowing through a square pipe can be visualized as a right rectangular prism. If water is flowing through a 2 in. × 2 in. square pipe at a flow speed of $4\frac{ft.}{s}$, then for every second the water flows through the pipe, the water travels a distance of 4 ft. The volume of water traveling each second can be thought of as a prism with a 2 in. × 2 in. base and a height of 4 ft. The volume of this prism is:

$$V = Bh$$
$$= \frac{1}{6} \text{ ft.} \times \frac{1}{6} \text{ ft.} \times 4 \text{ ft.}$$
$$= \frac{1}{9} \text{ ft}^3$$

Therefore, $\frac{1}{9}$ ft^3 of water flows every second, and the flow rate is $\frac{1}{9}\frac{\text{ft}^3}{\text{s}}$.

Problem Set

1. Harvey puts a container in the shape of a right rectangular prism under a spot in the roof that is leaking. Rainwater is dripping into the container at an average rate of 12 drops a minute. The container Harvey places under the leak has a length and width of 5 cm and a height of 10 cm. Assuming each raindrop is roughly 1 cm^3, approximately how long does Harvey have before the container overflows?

2. A large square pipe has inside dimensions 3 in. × 3 in., and a small square pipe has inside dimensions 1 in. × 1 in. Water travels through each of the pipes at the same constant flow speed. If the large pipe can fill a pool in 2 hours, how long will it take the small pipe to fill the same pool?

3. A pool contains 12,000 ft^3 of water and needs to be drained. At 8:00 a.m., a pump is turned on that drains water at a flow rate of 10 ft^3 per minute. Two hours later, at 10:00 a.m., a second pump is activated that drains water at a flow rate of 8 ft^3 per minute. At what time will the pool be empty?

4. In the previous problem, if water starts flowing into the pool at noon at a flow rate of 3 ft^3 per minute, how much longer will it take to drain the pool?

5. A pool contains 6,000 ft^3 of water. Pump A can drain the pool in 15 hours, Pump B can drain it in 12 hours, and Pump C can drain it in 10 hours. How long will it take all three pumps working together to drain the pool?

6. A 2,000-gallon fish aquarium can be filled by water flowing at a constant rate in 10 hours. When a decorative rock is placed in the aquarium, it can be filled in 9.5 hours. Find the volume of the rock in cubic feet (1 ft^3 = 7.5 gal.)

©2015 Great Minds. eureka-math.org
G7-M6-SE-B3-1.3.1-01.2016

Notes

Notes

Notes

Notes

Notes

Notes